HOW TO SOLVE

Algebra Problems Word

4th Edition

William A. Nardi

THOMSON

ARCO

Australia • Canada • Mexico • Singapore • Spain • United Kingdom • United States

An ARCO Book

ARCO is a registered trademark of Thomson Learning, Inc., and is used herein under license by Peterson's.

About The Thomson Corporation and Peterson's

With revenues of US$7.2 billion, The Thomson Corporation (www.thomson.com) is a leading global provider of integrated information solutions for business, education, and professional customers. Its Learning businesses and brands (www.thomsonlearning.com) serve the needs of individuals, learning institutions, and corporations with products and services for both traditional and distributed learning.

Peterson's, part of The Thomson Corporation, is one of the nation's most respected providers of lifelong learning online resources, software, reference guides, and books. The Education Supersite[SM] at www.petersons.com—the Internet's most heavily traveled education resource—has searchable databases and interactive tools for contacting U.S.-accredited institutions and programs. In addition, Peterson's serves more than 105 million education consumers annually.

For more information, contact Peterson's, 2000 Lenox Drive, Lawrenceville, NJ 08648; 800-338-3282; or find us on the World Wide Web at www.petersons.com/about.

Third Edition

COPYRIGHT © 2002 by William A. Nardi

Previous editions © 1999, 1991, 1987

ISBN 0-7689-1082-X

Printed in Canada

10 9 8 7 6 5 4 3 2   04 03 02

Contents

Acknowledgments

I wish to thank my daughter, Nadine Nardi Davidson, for her editorial assistance and faith in this project; Vernon Manley for his helpful suggestions; and my wife, Anne, for her patience and encouragement.

William A. Nardi, Sr.

Foreword

Word problems—they appear harmless at first, nothing more than a bunch of words, a couple of numbers, and some letters. But upon closer inspection, those words, numbers, and letters can leave you as clueless as Sherlock Holmes on a bad day. Unfortunately, every student, regardless of his or her field of study, will have to face algebra word problems over and over again. Algebra is required for high school math, it is necessary for advanced math such as calculus, and it appears on almost all of the major college and graduate level entrance exams such as the SAT, ACT, GRE, etc. Even if you have little difficulty solving algebra problems when the equation is given to you, you may find yourself continually frustrated or slowed down when these problems are presented in words only. Why? Because word problems add an extra challenge to the solving process—you first have to set up the equation by translating the word problem from English into an algebraic equation before it can be solved.

This book is a road map that will help you make your way through the word-problem maze by teaching you to recognize and isolate the key words in typical problems and then to translate them into algebraic equations.

Please note, however, that because algebra is the extension of arithmetic, it is assumed that all arithmetic fundamentals, including whole numbers, fractions, decimals, and percents, have already been mastered. You should also have a knowledge of solving equations with one and two unknowns and with positive and negative integers.

Chapter One

Translating English Terms Into Algebra Symbols

The most important and difficult problems for the algebra student to solve are word problems. These problems often use different words to mean the same thing. To make it easier to solve these problems, some frequently used words and their mathematical translations are listed below.

After you have studied this list, turn to page 5. The same list appears but without the translations. Write in the translations in the blank spaces. Do not go on to page 6 until you can identify all the terms.

TERM	SYMBOL OF TRANSLATION
add	+
sum	+
more	+
increased	+
greater	+
excess	+
less	−
difference	−
subtract	−
diminished	−
reduced	−
remainder	−
product	×
divide	÷
find quotient	÷

quantity	( )
equal	=
result	=
less than	<
greater than	>
greater than or equal to	≥
less than or equal to	≤

EXAMPLES:

WORDS	SYMBOL OF TRANSLATION
1. Nine less than twelve.	$12 - 9$
2. Four less than three times y.	$3y - 4$
3. Six less two.	$6 - 2$
4. Four times the quantity six plus two.	$4(6 + 2)$
5. One-half a number times fifteen.	$\frac{1}{2}n(15)$
6. One-third a number less two.	$\frac{1}{3}n - 2$
7. Ten times the sum of twice a number and six.	$10(2n + 6)$
8. Eighteen times the difference of a number and ten.	$18(n - 10)$
9. Eight times the quantity y plus two divided by four.	$\dfrac{8(y+2)}{4}$

TEST YOURSELF

Term	Symbol of Translation
add	_____
sum	_____
more	_____
increased	_____
greater	_____
excess	_____
less	_____
difference	_____
subtract	_____
diminished	_____
reduced	_____
remainder	_____
product	_____
divide	_____
find quotient	_____
quantity	_____
equal	_____
result	_____
less than	_____
greater than	_____
greater than or equal to	_____
less than or equal to	_____

TRANSLATING ENGLISH TERMS INTO ALGEBRA SYMBOLS

English	Algebra
(use *n* to represent the number)	
1. A number increased by six.	**1.** $n + 6$
2. Twice a number decreased by six.	**2.** $2n - 6$
3. Four less than five times a number.	**3.** $5n - 4$
4. The sum of three times a number and twelve.	**4.** $3n + 12$
5. One-half a number is fifty.	**5.** $\frac{1}{2}n = 50$
6. If from twice a number you subtract four, the difference is twenty.	**6.** $2n - 4 = 20$
7. Twenty-five is nine more than four times a number.	**7.** $25 = 9 + 4n$
8. Sixteen subtracted from five times a number equals the number plus four.	**8.** $5n - 16 = n + 4$
9. Twenty-five is the same as ten added to twice a number.	**9.** $25 = 10 + 2n$
10. Three *x* is five less than twice *x*.	**10.** $3x = 2x - 5$
11. The sum of two *y* and three is the same as the difference of three *y* and one.	**11.** $2y + 3 = 3y - 1$
12. The difference of five and five *y* is the same as eight and two *y*.	**12.** $5 - 5y = 8 + 2y$
13. Twice the quantity of two *y* and six.	**13.** $2(2y + 6)$
14. Twice the quantity of seven plus *x* is the same as the difference of *x* and seven.	**14.** $2(7 + x) = x - 7$
15. The sum of two *y* and the quantity of three plus *y* plus twice the quantity two *y* minus two equals fifteen.	**15.** $2y + (3 + y) + 2(2y - 2) = 15$

ENGLISH-ALGEBRA TRANSLATION TEST

Try translating the following from English to algebra (let *n* stand for the number). Then check your answers with those below. If you missed more than 3, restudy pages 1–4. and then try again.

ENGLISH	ALGEBRA
(use *n* to represent the number)	
1. A number added to five.	**1.** _____
2. The difference between six *n* and seven.	**2.** _____
3. One-half a number is thirty.	**3.** _____
4. Twice a number less twelve.	**4.** _____
5. Sixty-four diminished by *n* equals one.	**5.** _____
6. A number is equal to fifty less the number.	**6.** _____
7. Sixteen subtracted from five times a number equals the number plus four.	**7.** _____
8. Twice a number decreased by eight is zero.	**8.** _____
9. A number is equal to fifty less nine times the number.	**9.** _____
10. The product of twice a number and sixteen is sixty-four.	**10.** _____
11. If the sum of six and a number is divided by two, the quotient is twelve.	**11.** _____
12. If the product of five and a number is divided by six the result is ten.	**12.** _____
13. Twenty-five is five less than six *n* diminished by two.	**13.** _____
14. The quotient of sixteen *n* divided by one hundred equals sixteen.	**14.** _____

15. Twice the quantity of n + two

15. _____

divided by six is fifty.

ANSWERS TO ENGLISH-ALGEBRA TRANSLATION TEST

1. $n + 5$

2. $6n - 7$

3. $\frac{1}{2} n = 30$

4. $2n - 12$

5. $64 - n = 1$

6. $n = 50 - n$

7. $5n - 16 = n + 4$

8. $2n - 8 = 0$

9. $n = 50 - 9n$

10. $2(n + 16) = 64$

11. $\dfrac{(6+n)}{2} = 12$

12. $\dfrac{5n}{6} = 10$

13. $25 = (6n - 5) - 2$

14. $\dfrac{16n}{100} = 16$

15. $\dfrac{2(n+2)}{6} = 50$

Chapter Two

Translating English Statements Into Algebra Equations

Many word problems can be simplified by using the five hints listed below:

1. Read the problem at least twice.
2. Draw a diagram if possible. If it is a geometric problem, this is a *must*.
3. Underline the variable (unknown). When a problem concerns two things or two people, <u>whatever is mentioned last in the sentence is the unknown</u>.
4. Write an equation relating to the numbers in the problem. Let x represent the unknown or variable.

EXAMPLE 1:

John's age is twice <u>Mary's age</u>.

Let x = Mary's age (Mary's age was mentioned last.)
$\quad 2x$ = John's age

EXAMPLE 2:

(Always draw the linear figure.)

The length of a rectangle is 6 more than twice the <u>width</u> (mentioned last, so equals x).

Let x = width
$2x + 6$ = length

x

$2x + 6$

EXAMPLE 3:

There are 5 times as many dimes as <u>quarters</u>.

Let x = quarters
$\quad 5x$ = dimes

EXAMPLE 4:

The rate of a plane is 10 times that of a <u>freight train</u>.

Let x = rate of freight train
$\quad 10x$ = rate of plane

EXAMPLE 5:

Mr. Johnson invested 500 dollars more than twice <u>Mr. Rodriguez's investment</u>.

$\quad$ Let x = Mr. Rodriquez's investment
$500 + 2x$ = Mr. Johnson's investment

5. When you have 3 or more types of coins or other items in a problem, use x to represent whatever is <u>not mentioned</u>.

EXAMPLE 1:

Wilfred has 46 coins amounting to $6.50. The number of <u>dimes</u> is twice the number of nickels. The number of <u>quarters</u> is 6 more than the number of nickels. How many of each coin does he have?

$\quad$ Let x = number of nickels
$\quad 2x$ = number of dimes
$(x + 6)$ = number of quarters

EXAMPLE 2:

Harold has twice as much money as Nadine. Nadine has 3 times as much money as Sandra. If together they have $357, how much money does each of them have? (Notice that in the above problem, Sandra's money is not compared as stated with either Harold's or Nadine's. We therefore assign x to her amount.)

Let x = amount of money Sandra has

$3x$ = amount of money Nadine has

$6x$ = amount of money Harold has

Equation: $x + 3x + 6x = 357$

$$10x = 357$$
$$x = \$35.70 \text{ (Sandra)}$$
$$3x = \$107.10 \text{ (Nadine)}$$
$$6x = \$214.20 \text{ (Harold)}$$

Chapter Three

Solving Equations

Definition: When something equals something else, you have an equation.

EXAMPLES:

   a. $x = 6$

   b. $x + 5 = 6$

   c. $5n + 2n - 16 = 2n + 14$

   d. $6(n + 6) = -2(3n - 6) + n$

Some reminders:

Every equation has two members - a right and a left separated by an equal sign. When bringing (transposing) a term from a right member to a left member or from a left member to a right member, you must change the sign of the term from + (positive) to – (negative) or from – (negative) to + (positive).

Likewise, a term that is multiplied on one side is divided when moved or transposed to the opposite side, and a term that is divided on one side is multiplied on the opposite side.

To solve the equation, you first transpose like terms to the same side of the equation.

EXAMPLE 1:

$3x + 6 = 12$	1. Transpose like terms to same side of equation.
$3x = 12 - 6$	(When $+6$ is transposed, it becomes -6.)
$3x = 6$	2. Collect like terms ($12 - 6 = 6$.)
$x = 6 \div 3$	3. Transpose like terms.
	(When the 3 is transposed it is divided into the 6.)
$x = 2$	Solution.

EXAMPLE 2:

$8x + 16 = 5x + 1$ 1. Transpose like terms to same side of equation.

$8x - 5x = -16 + 1$ (Remember to change signs when crossing the equal sign.)

$3x = -15$ 2. Collect like terms.

$x = -15 \div 3$ 3. Transpose the 3 by dividing it into -15.

$x = -5$ Solution.

EXAMPLE 3:

$5(t-14) + 3(t+6) = -2(t+5) - 2$

$5t - 70 + 3t + 18 = -2t - 10 - 2$ 1. Remove parentheses.

$8t - 52 = -2t - 12$ 2. Collect like terms.

$8t + 2t = +52 - 12$ 3. Transpose like terms.

$10t = +40$ 4. Collect like terms.

$t = 40 \div 10$ 5. Transpose again.

$t = 4$ Solution

PRACTICE PROBLEMS: SOLVING EQUATIONS

1. The sum of two numbers is 231. The larger is twice the <u>smaller</u>. What are the numbers?

2. The length of a rectangle is three times the <u>width</u>. If the perimeter is 80 feet, what are the dimensions.

   (Reminder: The perimeter is the sum of all sides added together.)

   Remember to draw the figure.

3. The length of a rectangle is 6 less than twice the <u>width</u>. If the perimeter is 60 inches, what are the dimensions? (Draw a figure.)

4. A man earns $1000 more than 3 times what his <u>son earns</u>. If the total amount of what they both earn is $29,000, how much money did the father and son each earn?

5. Two tuna boats start from the same port at the same time, but they head in opposite directions. The faster boat travels 10 knots per hour faster than the <u>slower boat</u>. At the end of 8 hours, they were 272 nautical miles apart. How many nautical miles had each boat traveled by the end of the 8-hour period?

6.

The length of the larger square is twice as large as the <u>length of the smaller square</u>. If the sum of their perimeters is 648 centimeters, what is the length of each square?

7. A man's will left $75,000 to his wife, son, and daughter in such a way that his son's share was 3 times the <u>daughter's share</u>, while the wife received as much as the son and daughter together. How much did each receive?

Hint: Let x = daughter's share.

Check your answers with the solutions that follow.

ANSWERS TO PRACTICE PROBLEMS: SOLVING EQUATIONS

1. $5(t-14)+3(t+6)=-2(t+5)-2$

$5t-70+3t+18=-2t-10-2$	1. Remove parentheses.
$8t-52=-2t-12$	2. Collect like terms.
$8t+2t=+52-12$	3. Transpose like terms.
$10t=+40$	4. Collect like terms.
$t=40\div10$	5. Transpose again.
$t=4$	Solution

$2x+x=231$

$3x=231$

$x=77$ smaller number

$2x=154$ larger number

2.   Let x = width in feet

$3x$ = length in feet

x ft

3x ft

$2x+6x=80$ ft

$8x=80$

$x=10$ ft width

$3x=30$ ft length

3.   Let x = width in inches

$(2x-6)$ = length in inches

x in

2x − 6 in

$$2x + 2(2x - 6) = 60$$
$$2x + 4x - 12 = 60$$
$$6x = 72$$
$$x = 12 \text{ in width}$$
$$(2x - 6) = 18 \text{ in length}$$

4.   Let x = money son earns
$3x + 1,000$ = money father earns

$$x + 3x + 1,000 = 29,000$$
$$4x = 29,000 - 1,000$$
$$4x = 28,000$$
$$x = \$7,000 \text{ money son earns}$$
$$3x + 1,000 = \$22,000 \text{ money father earns}$$

5.   Let x = knots per hour for slower boat
$(x + 10)$ = knots per hour for faster boat
$8x$ = nautical miles traveled in 8 hrs by slower boat
$8(x + 10)$ = nautical miles traveled in 8 hrs by faster boat

$$8x + 8(x + 10) = 272$$
$$8x + 8x + 80 = 272$$
$$16x + 80 = 272$$
$$16x = 192$$
$$x = 12 \text{ knots per hour for slower boat}$$
$$x + 10 = 22 \text{ knots per hour for faster boat}$$
$$8x = 96 \text{ nautical miles traveled by slower boat}$$
$$8x + 80 = 176 \text{ nautical miles traveled by faster boat}$$

6. Let x = length of side in cm for smaller square

$4x$ = length of perimeter in cm of smaller square

Let $2x$ = length of side in cm for larger square

$8x$ = length of perimeter in cm of larger square

$$4x + 8x = 648 \text{ cm}$$
$$12x = 648 \text{ cm}$$
$$x = 54 \text{ cm, length of side for smaller square}$$
$$2x = 108 \text{ cm, length of side for larger square}$$
$$4x = 216 \text{ cm, length of perimeter of smaller square}$$
$$8x = 432 \text{ cm, length of perimeter of larger square}$$

7. Let x = money received by daughter

$3x$ = money received by son

$4x$ = money received by wife

$$x + 3x + 4x = \$75,000$$
$$8x = \$75,000$$
$$x = \$ \ 9,375 \text{ received by daughter}$$
$$3x = \$28,125 \text{ received by son}$$
$$4x = \$37,500 \text{ received by wife}$$

Chapter Four

Distance, Rate, and Time

A very important formula for solving these problems: $D = R \times T$

$$D = \text{Distance}$$
$$R = \text{Rate}$$
$$T = \text{Time}$$

Learn it.

Also, learn these important formulas: $\quad R = \dfrac{D}{T} \qquad T = \dfrac{D}{R}$

1. If a car travels at an average rate of 50 miles per hour, then in two hours it will have traveled __?__ miles.
2. If a plane traveled 2,184 miles in 3.5 hours, then its rate of speed is equal to __?__ miles per hour.
3. If a passenger train travels at an average rate of 55 miles per hour, then it will take __?__ hours to travel 275 miles.

When solving motion problems, always draw a diagram and then use the model illustrated in the example that follows.

EXAMPLE:

Two cars start from the same point and travel in opposite directions. The rate of the slower car is 15 miles per hour less than the rate of the faster car. After 8 hours they are 840 miles apart. Find the rate of speed of each car.

Diagram:

Let x = rate of faster car

$x - 15$ = rate of slower car

Model:

Formula	Distance	=	Rate	×	Hours
Faster car	$8x$		x		8
Slower car	$8(x-15)$		$x-15$		8

The distance of the faster car plus the distance of the slower car = Total Distance.

Equation:
$$8x+8(x-15)=840$$
$$8x+8x-120=840$$
$$16x=840+120$$
$$x=60 \text{ rate of faster car}$$
$$x-15=45 \text{ rate of slower car}$$

STUDY PROBLEM 1

Two trains leave the same train station at the same time, but in opposite directions. The faster train travels at an average rate of 75 miles per hour and the slower train travels at an average rate of 68 miles per hour. In how many hours will they be 715 miles apart?

Diagram:

Model:

Formula	Distance	=	Rate	×	Hours
Faster train	$75x$		75		x
Slower train	$68x$		68		x

Distance of faster train $= 75x$

Distance of slower train $= 68x$

Total distance $= 143x$

Equation:
$$143x=175$$
$$x=5 \text{ hours}$$

STUDY PROBLEM 2

Mr. Granger, a cyclist, rode from his home to his office at the average speed of 18 miles per hour. On his return home from his office, using the same route, he averaged 12 miles per hour. If the total round trip took 5 hours, what was the distance from his home to his office?

Diagram:

To office
Rate = 18 mph

Returning home
Rate = 12 mph

Model:

Formula	Distance	=	Rate	×	Hours
Home to office	$18x$		18		x
Office to home	$12(5-x)$		12		$5-x$

Equation:
$$18x = 12(5-x)$$
$$18x = 60 - 12x$$
$$18x + 12x = 60$$
$$30x = 60$$
$$x = 2 \text{ hours}$$
$$18x = 36 \text{ miles}$$

STUDY PROBLEM 3

Steve starts out on his moped at the rate of 40 miles per hour. Two hours later, Bill drives his car along the same route at 55 miles per hour. In how much time will Bill overtake Steve?

Diagram:

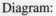
Bill's car

Steve's moped
2 hr at 40 mph

 Car 55 mph   Point where Bill will catch Steve.

Model:

Formula	Distance	=	Rate	×	Hours
Moped	$40(x + 2)$		40		$x + 2$
Car	$55x$		55		x

Equation:

$$55x = 40(x+2)$$
$$55x = 40x + 80$$
$$55x - 40x = 80$$
$$15x = 80$$
$$x = 5\tfrac{1}{3} \text{ hours} (5 \text{ hr and } 20 \text{ min})$$

PRACTICE PROBLEMS: DISTANCE, RATE, AND TIME

1. A 747 and a 727 leave the same airport at the same time and travel in opposite directions at 600 miles per hour and 450 miles per hour respectively. In how many hours will they be 3,150 miles apart?

Formula	Distance	=	Rate	×	Hours
?	?		?		?
?	?		?		?

Write equation. Answer is __?__ hours.

2. A helicopter traveling at an average rate of speed of 150 miles per hour left from a downtown terminal at noon after a train, which had departed from the same terminal at 9 a.m. If the helicopter overtook the train at 2:00 p.m., find the average speed of the train.

Draw diagram.

Draw model.

Write equation.

Answer is __?__ hours.

3. Two cars 720 kilometers apart travel toward each other. One car travels 70 kilometers per hour while the other car travels at the rate of 50 kilometers per hour. In how many hours will they meet?

Hint: Let x = hours it will take before they meet.

Distance covered by first car = 70 (the rate) × (time in hours).

Distance covered by second car = 50 (the rate) × (time in hours).

Equation: __?__

Answer is __?__ hours.

4. Two boys on bicycles start from the same place at the same time. One cyclist travels 15 miles per hour, the other travels 12 miles per hour. If they travel in the same direction, how many hours will it take before they are 9 miles apart?

Hint : Let x = time it takes to be nine miles apart.

$15x$ = distance of first cyclist.

$12x$ = distance of second cyclist.

Equation: __?__ .   Answer is __?__ hours.

5. A freight train takes 18 hours to travel the same distance that an express train travels in 15 hours. The rate of the express train is 15 miles per hour faster than the freight train. Find the rate at which each train travels.

> Draw diagram.
>
> Draw model.
>
> Write equation.
>
> Answer is _?_ mph for the express train.
>
> Answer is _?_ mph for the freight train.

6. At 10 a.m., 2 classmates start out on their bikes to meet each other from towns located 68 miles apart. At 1 p.m., they meet. If one boy traveled 3 miles per hour faster than the other, what was the rate of speed of each boy?

> Draw diagram.
>
> Draw model.
>
> Write equation.
>
> Answer is _?_ mph for the faster bike.
>
> Answer is _?_ mph for the slower bike.

7. At noon, two Mississippi steamboats are 126 miles apart. At 6 p.m., they pass each other going in opposite directions. If one steamboat travels 9 miles per hour faster than the other, find the speed of both boats.

> Draw diagram.
>
> Draw model.
>
> Write equation.
>
> Answer is _?_ mph for the faster boat.
>
> Answer is _?_ mph for the slower boat.

8. A jet bomber traveling 500 miles per hour arrived over its target at the same time as its fighter jet escort which left the same base $\frac{1}{2}$ hour after the bomber took off. How many hours did it take to reach the target if the fighter jet traveled 600 miles per hour?

> Draw diagram.
>
> Draw model.
>
> Write equation.
>
> Answer is _?_ hours.

9. A private jet had been flying for 2 hours when it encountered strong head winds which reduced its speed by 40 miles per hour. If it took the plane 5 hours to travel 1,380 miles, find its speed before flying into the head winds.

Hint: Let x = speed of jet during first 2 hours of flight.

Draw diagram.

Draw model.

Write equation.

Answer is __?__ mph.

Check your answers with the solutions that follow. If you missed more than 3, review this chapter again before going on to Chapter 5.

ANSWERS TO PRACTICE PROBLEMS: DISTANCE, RATE, AND TIME

1.

Formula	Distance	=	Rate	×	Time
747	600x		600		x
727	450x		450		x

Equation: $600x + 450x = 3{,}150$ miles (total distance)

$$1{,}050x = 3{,}150$$

$$x = 3 \text{ hr to be } 3{,}150 \text{ miles apart}$$

2.

Formula	Distance	=	Rate	×	Time
Train	5x		x		5
Helicopter	300		150		2

Equation: $5x = 300$

$$x = 60 \text{ mph is the average speed of the train}$$

3.

Formula	Distance	=	Rate	×	Time
Car 1	$70x$		70		x
Car 2	$50x$		50		x

Equation: $70x + 50x = 720$ km

$$120x = 720 \text{ km}$$

$$x = 6 \text{ hr before they meet}$$

4.

Formula	Distance	=	Rate	×	Time
1st boy	$15x$		15		x
2nd boy	$12x$		12		x

Equation: $15x - 12x = 9$

$$3x = 9$$

$$x = 3 \text{ hr}$$

5.

Formula	Distance	=	Rate	×	Time
Freight train	$18x$		x		18 hr
Express train	$15(x + 15)$		$x + 15$		15 hr

Equation:
$$18x = 15(x+15)$$
$$18x = 15x + 225$$
$$18x - 15x = 225$$
$$3x = 225$$
$$x = 75 \text{ mph for the freight train}$$
$$x + 15 = 90 \text{ mph for the express train}$$

6.

Formula	Distance	=	Rate	×	Time
Faster cyclist	$3(x + 3)$		$x + 3$		3 hr
Slower cyclist	$3x$		x		3 hr

d_1 = distance of faster cyclist

d_2 = distance of slower cyclist

$d_1 + d_2$ = total distance = 68 miles

Equation:
$$3(x+3)+3x = 68$$
$$3x + 9 + 3x = 68$$
$$6x = 59$$
$$x = 9\,\tfrac{5}{6} \text{ mph for the slower cyclist}$$
$$x + 3 = 12\,\tfrac{5}{6} \text{ mph for the faster cyclist}$$

7.

Formula	Distance	=	Rate	×	Time
1st steamboat	$6x$		x		6 hr
2nd steamboat	$6(x + 9)$		$x + 9$		6 hr

$d_1 + d_2$ = total distance covered by the 2 boats

Equation:
$$6x + 6(x+9) = 126$$
$$6x + 6x + 54 = 126$$
$$12x = 72$$
$$x = 6 \text{ mph for first boat}$$
$$x + 9 = 15 \text{ mph for second boat}$$

8.

Formula	Distance	=	Rate	×	Time
Jet bomber	$500(x + .5)$		500		$x + .5$
Fighter jet	$600x$		600		x

Equation:

$$600x = 500(x + .5)$$
$$600x = 500x + 250$$
$$600x - 500x = 250$$
$$100x = 250$$
$$x = 2\tfrac{1}{2} \text{ hr for fighter jet}$$
$$x + \tfrac{1}{2} = 3 \text{ hr for jet bomber}$$

9.

Formula	Distance	=	Rate	×	Time
Before headwind	$2x$		x		2 hr
With headwind	$3(x - 40)$		$x - 40$		3 hr

Equation:
$$2x + 3(x - 40) = 1,380 \text{ miles}$$
$$2x + 3x - 120 = 1,380$$
$$5x - 120 = 1,380$$
$$5x = 1,380 + 120$$
$$5x = 1,500$$
$$x = 300 \text{ mph speed before flying into headwind}$$

Chapter Five

Coin Problems

When solving coin problems, all coins must be changed to cents.

EXAMPLES: (2) $3x + 3 = 9$

$$3x = 6$$

$$x = 2$$

Since there are 100 cents in 1 dollar, then there are 540 cents in $5.40. In $11.45, there would be 1,145 cents.

STUDY PROBLEM 1

A collection of quarters and nickels contains twice as many quarters as <u>nickels</u>. If the value of the coins totals $4.40, how many nickels and quarters are in the collection?

Let x = number of nickels

$2x$ = number of quarters

$5(x)$ = value in cents of the number of nickels

$25(2x)$ = value in cents of the number of quarters

$4.40 = 440$ cents

Equation: $5x + 50x = 440$ (all terms in cents)

$$55x = 440$$

$$x = 8 \text{ (number of nickels)}$$

$$2x = 16 \text{ (number of quarters)}$$

Use the format in the example on page 23 to solve the following problems.

PRACTICE PROBLEMS: COINS

(Check your answers with the solutions on pages 32–35.)

1. In a collection of coins that has a value of $4.90, there are 20 more nickels than dimes. How many nickels and dimes are there in the collection?

$$\text{Let } x = \text{number of } \underline{?}$$
$$\underline{?} = \text{number of } \underline{?}$$
$$10(x) = \text{value in cents of } \underline{?}$$
$$5(20+x) = \text{value in cents of } \underline{?}$$
$$\$4.90 = \underline{?} \text{ cents}$$

Equation: __?__ .

a. Number of dimes is __?__

b. Number of nickels is __?__

2. A student at Ellwood High has a collection of quarters and nickels amounting to $18.70. If there are 20 more nickels than quarters, how many of each kind of coin does the student have?

$$\text{Let } x = \text{number of } \underline{?}$$
$$\underline{?} = \text{number of } \underline{?}$$
$$25(x) = \text{value in cents of } \underline{?}$$
$$5(x+20) = \text{value in cents of } \underline{?}$$
$$\$18.70 = \underline{?} \text{ cents}$$

Equation: __?__ .

a. Number of quarters is __?__

b. Number of nickels is __?__

3. Sylvia has $11.25 in nickels, dimes, and quarters. She has 3 times as many nickels as dimes and 5 more quarters than dimes. How many of each kind of coin does she have?

$$\text{Let } x = \text{number of \_\_?\_\_}$$
$$3x = \text{number of \_\_?\_\_}$$
$$\underline{\text{(?)}} = \text{number of \_\_?\_\_}$$
$$\underline{\_?\_} = \text{value in cents of dimes}$$
$$\underline{\_?\_} = \text{value in cents of nickels}$$
$$\underline{\_?\_} = \text{value in cents of quarters}$$
$$\$11.25 = \underline{\_?\_} \text{ cents}$$

Equation: _?_ .

a. Number of dimes is _?_

b. Number of nickels is _?_

c. Number of quarters is _?_

4. A newsboy collected $68.50 on his paper route one week. The collection consisted of nickels, dimes, and quarters. If he had 70 more nickels than quarters and 20 more dimes than nickels, how many of each kind of coin did he have?

$$\text{Let } x = \text{number of \_\_?\_\_}$$
$$\underline{\_?\_} = \text{number of nickels}$$
$$\underline{\_?\_} = \text{number of dimes}$$
$$25(x) = \text{value in cents of quarters}$$
$$\underline{\_?\_} = \text{value in cents of nickels}$$
$$\underline{\_?\_} = \text{value in cents of dimess}$$
$$\$68.50 = \underline{\_?\_} \text{ cents}$$

Equation: _?_ .

a. Number of quarters is _?_

b. Number of nickels is _?_

c. Number of dimes is _?_

Whenever the *total number* of coins, tickets, or stamps is given in a problem, follow the reasoning below.

If you have a stack of 22 coins made up of nickels and dimes and you are told that 12 of the coins are dimes, then 22 – 12 = 10 nickels.

If you know that 7 of the coins are nickels, then 22 – 7 = 15 dimes.

If you have x number of dimes, then 22 – x = the number of nickels.

STUDY PROBLEM 2

A purse contained 58 coins consisting of dimes and nickels. If the total amount of these coins amounted to $4.80, how many of each kind of coin are in the purse?

$$\text{Let } x = \text{number of dimes}$$
$$58 - x = \text{number of nickels}$$
$$10(x) = \text{value in cents of } x \text{ dimes}$$
$$5(58 - x) = \text{value in cents of } (58 - x) \text{ nickels}$$
$$\$4.80 = 480 \text{ cents}$$

Equation: $10x + 5(58 - x) = 480$
$$10x + 290 - 5x = 480$$
$$5x + 290 = 480$$
$$5x = 190$$
$$x = 38 \text{ dimes}$$
$$58 - x = 20 \text{ nickels}$$

MORE PRACTICE PROBLEMS

(Check your answers with the solutions on pages 32–35.)

6. Eighty-six coins made up of dimes and quarters amount to $14.60. How many of each of these coins are there?

$$\text{Let } x = \text{number of quarters}$$
$$\underline{\quad(?)\quad} = \text{number of dimes}$$
$$\underline{\quad?\quad} = \text{value in cents of quarters}$$
$$\underline{\quad?\quad} = \text{value in cents of dimes}$$
$$\$14.60 = \underline{\quad?\quad} \text{ cents}$$

Equation: _?_.

a. Number of quarters is _?_

b. Number of dimes is _?_

7. A piggy bank contained 110 coins including nickels, dimes, and quarters. If there are twenty more nickels than quarters, and the number of dimes is equal to the number of quarters, how many of each kind of coin are there if the total collection is $13?

$$\text{Let } x = \text{number of } \underline{\quad?\quad}$$
$$x = \text{number of } \underline{\quad?\quad}$$
$$\underline{\quad(?)\quad} = \text{number of nickels}$$
$$10x = \text{value in cents of } \underline{\quad?\quad}$$
$$?x = \text{value in cents of quarters}$$
$$5(20+x) = \text{value in cents of } \underline{\quad?\quad}$$
$$\$13 = \underline{\quad?\quad} \text{ cents}$$

Equation: _?_.

a. Number of dimes is _?_

b. Number of quarters is _?_

c. Number of nickels is _?_

8. A high school play grossed $122.50. If there were twice as many student tickets sold as there were adult tickets, and if each student ticket sold for 50 cents and each adult ticket for 75 cents, how many student and how many adult tickets were sold for the play?

(Follow the same procedure as in solving coin problems.)

    a.   Adult tickets sold = _?_

    b.   Student tickets sold = _?_

ANSWERS TO PRACTICE PROBLEMS: COINS

1.

$$x = \text{number of } \underline{\text{dimes}}$$
$$x + 20 = \text{number of } \underline{\text{nickels}}$$
$$10(x) = \text{value in cents of } \underline{\text{dimes}}$$
$$5(20 + x) = \text{value in cents of } \underline{\text{nickels}}$$
$$\$4.90 = \underline{490} \text{ cents}$$

Equation:
$$10x + 5(20 + x) = 490$$
$$10x + 100 + 5x = 490$$
$$15x = 490 - 100$$
$$15x = 390$$
$$x = \underline{26 \text{ dimes}}$$
$$20 + x = \underline{46 \text{ nickels}}$$

2.

$$x = \text{number of } \underline{\text{quarters}}$$
$$x + 20 = \text{number of } \underline{\text{nickels}}$$
$$25(x) = \text{value in cents of } \underline{\text{quarters}}$$
$$5(x + 20) = \text{value in cents of } \underline{\text{nickels}}$$
$$\$18.70 = \underline{1,870 \text{ cents}}$$

Equation: $25x + 5(x + 20) = 1,870$

$$25x + 5x + 100 = 1,870$$

$$30x + 100 = 1,870$$

$$30x = 1,870 - 100$$

$$30x = 1,770$$

$$x = \underline{59 \text{ quarters}}$$

$$x + 20 = \underline{79 \text{ nickels}}$$

3.

$x = $ number of <u>dimes</u>

$3x = $ number of <u>nickels</u>

$\underline{x + 5} = $ number of <u>quarters</u>

$\underline{10(x)} = $ value in cents of dimes

$\underline{5(3x)} = $ value in cents of nickels

$\underline{25(x + 5)} = $ value in cents of quarters

$\$11.25 = \underline{1,125}$ cents

Equation: $10(x) + 5(3x) + 25(x + 5) = 1,125$

$$10x + 15x + 25x + 125 = 1,125$$

$$50x + 125 = 1,125$$

$$50x = 1,125 - 125$$

$$50x = 1,000$$

$$x = \underline{20 \text{ dimes}}$$

$$3x = \underline{60 \text{ nickels}}$$

$$(x + 5) = \underline{25 \text{ quarters}}$$

4.   Let x = number of quarters

$\underline{70 + x}$ = number of nickels

$\underline{90 + x}$ = number of dimes

$\underline{25(x)}$ = value in cents of quarters

$\underline{5(70 + x)}$ = value in cents of nickels

$\underline{10(90 + x)}$ = value in cents of dimes

$\$68.50 = \underline{6,850}$ cents

Equation: $\cos = \dfrac{adj.\ leg}{Hyp}$

$\cos = \dfrac{17.2}{AB}$ (hypotenus)

5.   Let x = number of quarters

$\underline{86 - x}$ = number of dimes

$\underline{25(x)}$ = value in cents of quarters

$\underline{10(86 - x)}$ = value in cents of dimes

$\$14.60 = \underline{1,460}$ cents

Equation: $25(x) + 10(86 - x) = 1,460$

$25x + 860 - 10x = 1,460$

$15x = 1,460 - 860$

$15x = 600$

$x = \underline{40}$ quarters

$86 - x = \underline{46}$ dimes

6. x = number of quarters

             x = number of dimes

        $20 + x$ = number of nickels

          $10x$ = value in cents of dimes

          $25x$ = value in cents of quarters

      $5(20 + x)$ = value in cents of nickels

         $\$13 = 1{,}300$ cents

Equation: $25x + 10x + 5(20 + x) = 1{,}300$

$$25x + 10x + 100 + 5x = 1{,}300$$

$$40x = 1{,}300 - 100$$

$$40x = 1{,}200$$

$$x = 30 \text{ dimes}$$

$$x = 30 \text{ quarters}$$

$$20 + x = 50 \text{ nickels}$$

7.   Let x = number of adult tickets sold

          $2x$ = number of student tickets sold

      $75(x)$ = value in cents of the adult tickets sold

     $50(2x)$ = value in cents of student tickets sold

   $\$122.50 = 12{,}250$ cents

Equation: $75x + 100x = 12{,}250$

$$175x = 12{,}250$$

$$x = 70 \text{ adult tickets sold}$$

$$2x = 140 \text{ student tickets sold}$$

Chapter Six

Mixture Problems: Review of Percent

Without adequate knowledge of percents, mixture problems will be impossible for you to solve. Take the review test below without looking at the answers on the next page. Study the answers that you missed. If you do not understand, go back and review the test before going on.

PART I

Percent means hundredths. Change the following percents to decimal fractions.

Example: 5% is .05

1. 3%
2. 4%
3. 8%
4. 11%
5. 22%
6. 37%
7. 44%
8. 59%
9. 46%
10. 88%

11. $5\frac{1}{2}\%$
12. $6\frac{1}{2}\%$
13. $8\frac{1}{2}\%$
14. $12\frac{1}{2}\%$
15. 155%
16. 275%
17. $8\frac{1}{4}\%$

18. $9\frac{3}{4}\%$
19. $10\frac{3}{4}\%$
20. $112\frac{1}{2}\%$
21. 6.8%
22. 12.8%
23. 127.3%
24. 2.8%
25. .5%

PART II

Change the percents below to decimals before multiplying.

1. 5% of 250 = ?

2. 7% of 336 = ?

3. 32% of 12,500 = ?

4. 6% of x = ?

5. 6% of (10,000 − x) = ?

6. $12 \frac{1}{4}$% of 800 = ?

7. 6.5% of $44,500 = ?

8. 300% of 50 = ?

9. $8 \frac{1}{2}$% of $6,180 = ?

10. .8% of 42,000 = ?

Check your answers before going on.

ANSWERS TO REVIEW OF PERCENT

PART I

1. .03

2. .04

3. .08

4. .11

5. .22

6. .37

7. .44

8. .59

9. .46

10. .88

11. $.05 \frac{1}{2}$ or .055

12. $.06 \frac{1}{2}$ or .065

13. $.08 \frac{1}{2}$ or .085

14. $.12 \frac{1}{2}$ or .125

15. 1.55

16. 2.75

17. $.08 \frac{1}{4}$ or .0825

18. $.09 \frac{3}{4}$ or .0975

19. $.10 \frac{3}{4}$ or .1075

20. $1.12 \frac{1}{2}$ or 1.125

21. .068

22. .128

23. 1.273

24. .028

25. .005

PART II

1. 12.50	6. 98
2. 23.52	7. $2,892.50
3. 4,000	8. 150
4. .06x	9. $525.30
5. 600 – .06x	10. $336

When solving mixture problems, there will be 3 different percentages of 3 different costs per pound of candy, nuts, fruits, or other product. When the problem involves chemicals or other liquids, a liquid measurement such as ounces, grams, cubic centimeters, liters, quarts, gallons, or pints will be given. We will use the model below to help us find the solution to mixture problems.

Model:

1st solution	2nd solution	mixture
?%	?%	?%

STUDY PROBLEM 1

How many ounces of pure alcohol must be added to a 42-oz solution containing 10% alcohol to make a solution containing 40% alcohol?

Model:

original solution	pure alcohol	mixture (new solution)
42 oz	x oz	42 + x oz
10%	100%	40%

Change each percent to a decimal fraction as follows:

original solution   pure alcohol   mixture (new solution)

42 oz   +   x oz   =   42 + x oz

.10   1.00   .40

Multiply each decimal fraction by the contents in each container:

Equation: $.10(42)+1.00(x)=.40(42+x)$

$$4.20+1.00x=16.80+.40x$$
$$100x-40x=1,680-420$$
$$60x=1,260$$
$$x=21 \text{ oz of pure alcohol must be added.}$$

The new mixture will contain 63 oz.

STEPS FOR SOLVING MIXTURE PROBLEMS

1. Draw model.

2. Place percents outside containers.

3. Change percents to decimals (hundredths).

4. Multiply each decimal fraction by the contents in each container.

5. Write equation.

6. Solve equation.

STUDY PROBLEM 2

How many liters of a 10% acid solution must be added to 2 liters of a 25% acid solution to make a mixture containing a 20% acid solution?

Model:

1st solution

x
liters
of acid

$10\% = .10$

+

2nd solution

2
liters
of acid

$25\% = .25$

=

mixture

$x + 2$
liters
of acid

$20\% = .20$

Equation: $.10x + .25(2) = .20(x + 2)$

$.10x + .50 = .20x + .40$

$$50 - 40 = 20x - 10x$$
$$10 = 10x$$
$$1 \text{ liter} = x \,(\text{answer})$$

Note: When using water to reduce the percentage of the solution, use 0% on the outside of the container that represents the water.

STUDY PROBLEM 3

To reduce 32 grams of a 25% solution of antiseptic to a 20% solution, how much distilled water should a pharmacist add?

32
grams

25%

+

x grams
water

0%

=

$x + 32$
grams

20%

(Change percent to decimal)

.25   0   .20

Equation: $.25(32)+0=.20(x+32)$

$8.00+0=.20x+6.40$

$800-640=20x$

$160=20x$

$8=x$ grams of water to be added.

PRACTICE PROBLEMS: MIXTURES

(Check your answers with the solutions on pages 47–54.)

1. To reduce 32 grams of a 25% solution of antiseptic to a 10% solution, how much distilled water should a pharmacist add?

acid mixture   distilled water   new mixture

32 grams   +   x grams of water   =   x + 32 grams

25%   0%   10%

a. Equation: __?__ .

b. Answer is __?__ .

2. How many liters of a 20% solution of acid should be added to 10 liters of a 30% solution of acid to obtain a 25% solution?

a. Equation: _?_ .

b. Answer is _?_ .

3. How many quarts of a 50% solution of acid must be added to 20 quarts of a 20% solution of acid to obtain a mixture containing a 40% solution of acid?

a. Equation: _?_ .

b. Answer is _?_ .

4. A chemist has a solution of 50% pure acid and another of 80% pure acid. How many gallons of each will make 600 gallons of 72% pure acid?

a. Equation: _?_ .

b. Answer is _?_ gallons for 50% pure acid.

c. Answer is _?_ gallons for 80% pure acid.

5. A storekeeper wishes to sell 100 pounds of mixed nuts at $1.75 a pound. He mixes peanuts worth $1.65 a pound with cashews worth $1.90 a pound. How many pounds of each does he use?

$1.90 $1.65 $1.75

   a. Equation: __?__ .

   b. __?__ lb at $1.65.

   c. __?__ lb at $1.90.

6. A merchant wishes to blend 200 pounds of coffee worth $3.20 a pound from two mixtures—one at $3 and the other at $3.25. How many pounds of each mixture should he use?

7. A dairy wishes to mix 1,000 pounds of milk containing 8% butterfat. If the mixture is to be made from milk containing 5% butterfat and cream containing 20% butterfat, how many pounds of each is needed?

   a. Equation: __?__ .

   b. Answer in pounds at 5% is __?__ .

   c. Answer in pounds at 20% is __?__ .

8. How many ounces of an alloy containing 30% gold on the world market must be mixed with an alloy containing 5% gold to obtain 25 ounces of an alloy containing 20% gold?

9. A pharmacist has a 12% solution and a 20% solution of boric acid. How much of each must he use to make 80 grams of a 15% solution?

10. A druggist has an 18% solution of argyol. How much of this solution and how much water must be mixed together to make 10 liters of a 12% solution?

11. A grocer wishes to blend two different coffee brands. He wants to make a blend of 480 pounds to sell at $2.68 a pound. If he uses a blend of coffee worth $2.50 a pound with another blend worth $2.80 a pound, how many pounds of each does he use?

12. The Patronete winery's tastiest wine must contain a 12% alcohol content. How many gallons of wine with a 9% alcohol content must be mixed with 3,000 gallons of wine with a 15% alcohol content in order to achieve the desired 12% alcohol content?

13. The Davidson winery is noted for its Regal champagne. It takes a blending of 1,000 bushels of a mixture of two varieties of grapes to achieve the Regal specification of 12% alcohol content for its champagne. One of the two varieties of grape contains a 10% alcohol content and the other contains a 15% alcohol content. How many bushels of each variety are needed?

Note: In some mixture problems you not only add something to the mixture, but you also drain off some of the mixture. Four containers will be used in the following format.

STUDY PROBLEM 4

A 7-quart truck radiator was assumed to contain an 80% rating in alcohol, but after checking, it was found to contain a rating of only 60% alcohol. How much 90% alcohol must be added and how much 60% alcohol must be drained off to achieve an 80% rating?

   Model:

$$.90\% x + .60\%(7) - .60x = .80\%(7 \text{ qts})$$

Move decimal point 2 places to the right

$$.\underline{90}x + .\underline{60}(7) - .\underline{60}\% x = .\underline{80}(7)$$

$$90x + 420 - 60x = 560$$

$$90x - 60x = 560 - 420$$

$$30x = 140$$

$$x = 4\frac{2}{3} \text{ qts}$$

must be added and drained off

MORE PRACTICE PROBLEMS

(Check your answers with the solutions on page 47–54.)

14. The capacity of a car radiator is 20 quarts. If it's full of a 55% antifreeze solution, how many quarts must be drained off and replaced by a 80% solution to give 20 quarts of a 70% solution?

15. The capacity of a car radiator is 18 quarts. If it's full of a 20% antifreeze solution, how many quarts must be drained off and replaced by a 100% solution to give 18 quarts of a 39% solution?

ANSWERS TO PRACTICE PROBLEMS: MIXTURES

1. a. $.25(32) + 0\%(x) = .10(x + 32)$

Move decimal point 2 places to the right

$$25(32) + 0x = 10(x + 32)$$
$$800 + 0 = 10x + 320$$
$$800 - 320 = 10x$$
$$480 = 10x$$

b.   48 grams of water $= x$ to be added

2. a. $.20(x) + .30(10) = .25(x + 10)$

Move decimal point 2 places to the right

$$20x + 300 = 25x + 250$$
$$300 - 250 = 25x - 20x$$
$$50 = 5x$$

b.   10 liters of 20% acid $= x$

3. a. $50x + .20(20) = .40(x + 20)$

   Move decimal point 2 places to the right

$$50x + 400 = 40x + 800$$
$$50x - 40x = 800 - 400$$
$$10x = 400$$

b. $x = 40$ quarts to be added

4. a. $50(x) + .80(600 - x) = .72(600)$

   Move decimal point 2 places to the right

$$50x + 48,000 - 80x = 43,200$$
$$50x - 80x = -48,000 + 43,200$$
$$-30x = -4,800$$

b. $x = 160$ gal @ 50%

c. $600 - x = 440$ gal @ 80%

5. $x = $ lb of peanuts

   $(100 - x) = $ lb of cashews

a. $1.90(100 - x) + 1.65(x) = 1.75(100)$

   Move decimal point 2 places to the right

$$19,000 + 190x + 165x = 17,500$$
$$-25x = 17,500 - 19,000$$
$$-25x = -1,500$$
$$x = 60 \text{ lb peanuts @ \$1.65}$$
$$100 - x = 40 \text{ lb cashews @ \$1.90}$$

(*Note* : A negative divided by a negative equals a positive.)

b. $x = 60$ lb of peanut

c. $100 - x = 40$ lb of cashews

6.

Equation: $3.00(x) + 3.25(200 - x) = 3.20(200)$

Move decimal point 2 places to the right

$$300x + 65,000 - 325x = 64,000$$
$$300x - 325x = 64,000 - 65,000$$
$$-25x = -1,000$$

(*Note* : A negative divided by a negative equals a positive.)

$$x = 40 \text{ lb @ \$3.00}$$
$$200 - x = 160 \text{ lb @ \$3.25}$$

7.

a. $.05(x) + .20(1,000 - x) = .08(1,000)$

Move decimal point 2 places to the right

$$5x + 20,000 - 20x = 8,000$$
$$5x - 20x = 8,000 - 20,000$$
$$-15x = -12,000$$

b. $x = 800 \text{ lb @ 5\%}$

c. $100 - x = 200 \text{ lb @ 20\%}$

8.

Equation: $.30(x)+.05(25-x)=.20(25)$

Move decimal point 2 places to the right

$$30x+125-5x=500$$
$$30x-5x=500-125$$
$$25x=375$$
$$x=15 \text{ oz of a } 30\% \text{ alloy}$$
$$25-x=10 \text{ oz of a } 5\% \text{ alloy}$$

9.

Equation: $.12(x)+.20(80-x)=.15(80)$

Move decimal point 2 places to the right

$$12x+1,600-20x=1,200$$
$$-8x=1,200-1,600$$
$$-8x=-400$$
$$x=50 \text{ grams of } 12\% \text{ solution}$$
$$80-x=30 \text{ grams of } 20\% \text{ solution}$$

10.

Equation: $.1\underset{\smile}{8}(x) + 0\%(10 - x) = .1\underset{\smile}{2}(10)$

Move decimal point 2 places to the right

$$18x + 0 = 120$$
$$18x = 120$$
$$x = 6\frac{2}{3} \text{ liters of 18\% solution}$$
$$(10 - x) = 3\frac{1}{3} \text{ liters of water}$$

11.

Equation: $2.\underset{\smile}{50}(x) + 2.\underset{\smile}{80}(480 - x) = 2.\underset{\smile}{68}(480)$

Move decimal point 2 places to the right

$$250x + 134,000 - 280x = 128,640$$
$$-30x = 128,640 - 134,400$$
$$-30x = -5,760$$

(*Note*: A negative divided by a negative equals a positive.)

$$x = 192 \text{ lb @ \$2.50 per lb}$$
$$480 - x = 288 \text{ lb @ \$2.80 per lb}$$

12.

Equation: $.09(x)+.15(3,000)=.12(3,000+x)$

Move decimal point 2 places to the right

$$9x+45,000=36,000+12x$$
$$45,000-36,000=12x-9x$$
$$9,000=3x$$
$$x=3,000 \text{ gal of 9\% alcohol}$$

content needed

13. Equation: $.10(x)+.15(1,000-x)=.12(1,000)$

Move decimal point 2 places to the right

$$10x+15,000-15x=12,000$$
$$-5x=-3,000$$
$$x=600 \text{ bushels of the 10\% variety}$$
$$1,000-x=400 \text{ bushels of the 15\% variety}$$

14.

$$I - II + III = IV$$
$$.\underset{\smile}{55}(20) - .\underset{\smile}{55}(x) + .\underset{\smile}{80}(x) = .\underset{\smile}{70}(20)$$

Move decimal point 2 places to the right

$$1,100 - 55x + 80x = 1,400$$
$$80x - 55x = 1,400 - 1,100$$
$$25x = 300$$
$$x = 12 \text{ quarts}$$

must be drained off and replaced by an 80% solution

15.

$$I - II + III = IV$$

$$.\underset{\smile}{20}(18) - .\underset{\smile}{20}(x) + 1.\underset{\smile}{00}(x) = .\underset{\smile}{39}(18)$$

Move decimal point 2 places to the right

$$360 - 20x + 100x = 702$$

$$80x = 342$$

$$x = 4\frac{11}{40}$$

quarts of a 20% solution must be drained off and replaced by a 100% solution

Chapter Seven

Investment Problems

Definition of terms used in the investment business:

1. Financial instruments:

   a. Common stocks

   b. Corporate bonds

   c. Real estate mortgages

   d. Savings accounts

   e. U.S. Treasury Notes

2. Yield: Interest or dividends earned or income from a particular financial instrument. All the instruments in paragraph 1 earn interest or dividends.

   This model represents any financial instrument. Use it to solve investment problems.

   Model:

$$\boxed{\begin{matrix}\text{Dollars} \\ \text{invested}\end{matrix}} \; + \; \boxed{\begin{matrix}\text{Dollars} \\ \text{invested}\end{matrix}} \; = \; \boxed{\begin{matrix}\text{Yield} \\ (\text{Money earned})\end{matrix}}$$

   Rate of   Rate of

   interest   interest

STUDY PROBLEM 1

A man wishes to invest $10,000. He places a certain amount in a bank which pays 6% interest and the remainder in a stock which yields a 7% dividend. How much money did he invest in the bank and how much in stock if his total earnings for 1 year are $665.00?

	Bank		Stock		Annual
	Dollars	+	Dollars	=	Yield
	x		$\$10,000 - x$		$\$665.00$
	6%=.06		7% = .07		

Equation: $.06x + .07(\$10,000 - x) = \$665.00$

Move decimal point 2 places to the right

$$6x + \$70,000 - 7x = \$66,500$$
$$-x = \$66,500 - \$70,000$$
$$-x = -\$3,500\,^*$$
$$x = \$3,500 \text{ invested in the bank}$$
$$\$10,000 - x = \$6,500 \text{ invested in stocks}$$

*Reminder: A negative divided by a negative becomes a positive.

STUDY PROBLEM 2

Mr. Kim, a diamond dealer, invested part of his $10,000 in a U.S. Treasury Note at 8½% and the remainder in a corporate bond at 10½%. If the income from both of these investments yielded $970, how much did he invest at each rate?

	Treasury Note		Corporate Bond		Annual Yield
	x	+	$\$10,000 - x$	=	$\$970.00$
	$8\frac{1}{2}\%$		$10\frac{1}{2}\%$		

Equation: $.08\frac{1}{2}(x) + .10\frac{1}{2}(10,000 - x) = 970.00$

Move decimal point 2 places to the right

$$8\frac{1}{2}x + 105,000 - 10\frac{1}{2}x = 97,000$$
$$-2x = -8,000$$
$$x = \$4,000 \text{ invested at } 8\frac{1}{2}\%$$
$$10,000 - x = \$6,000 \text{ invested at } 10\frac{1}{2}\%$$

STUDY PROBLEM 3

The income from the first investment is $500.00 more than the income from the second investment. If $1000 more is invested in the first investment which pays 9% than in the second investment paying 8%, how much was invested in each?

1st Investment		2nd Investment		Annual Yield
$x+\$1,000$	$-$	x	$=$	$\$500.00$
9%		8%		

Equation: $.09(x+1,000)-.08(x)=500.00$

Move decimal point 2 places to the right

$$9x+9,000-8x=50,000$$
$$9x-8x=50,000-9,000$$
$$x=\$41,000 \text{ invested at } 8\%$$
$$x+1,000=\$42,000 \text{ invested at } 9\%$$

Now solve these problems. Check your answers with those on pages 59–63.

PRACTICE PROBLEMS: INVESTMENT

1. On her 16th birthday, Jane received $1000 from her grandfather. If she invested the money in 2 banks, one paying 6% and the other 7%, how much did she invest in each bank if her annual interest was $65?

1st Bank		2nd Bank		Annual Yield
x	$+$	$?$	$=$	$\$65.00$
6%		7%		

Equation: _?_.
Answers are _?_ and _?_.

2. Part of $12,000 is invested in 9% bonds and the remainder in common stocks paying 8%. If the yearly income from both investments is $1040, how much was invested in bonds and how much was invested in stocks?

3. Mrs. Walsh has $14,000 to invest. She chooses two investments, one paying $8\frac{1}{2}\%$ and one paying $7\frac{1}{2}\%$. If her yearly earnings from both investments is $1152, what did she invest at $8\frac{1}{2}\%$?

4. Mr. Camacho invested twice as much in a savings account at $6\frac{1}{2}\%$ as he invested in a U.S. Treasury Note at 7%. If his yearly income from both investments is $800, how much has he invested at $6\frac{1}{2}\%$ and how much at 7%?

Savings Account		Treasury Note		Annual Yield
?	+	x	=	$800.00
$6\frac{1}{2}\%$		7%		

Equation: _?_.
Amount invested at 6½ is _?_.
Amount invested at 7% is _?_.

5. Mrs. Chin, on reaching age 65, invested $18,000 of her insurance policy in AA bonds paying $8\frac{1}{2}\%$ and a common stock paying $7\frac{1}{2}\%$. If her yearly income from these 2 investments is $1470, how much did she invest in the bonds and how much in the stock?

6. A man had $12,000 to invest. If his income from a savings account earning $7\frac{1}{2}\%$ is $280 greater than his income from another investment at 8%, how much did he invest at each rate?

7. Two sums of money differing by $2000 were invested, the larger amount at $8\frac{1}{2}\%$ and the smaller amount at $9\frac{1}{2}\%$. If the larger sum yielded an annual income that was $110 more than the other, how much money was invested at each rate?

8. Mrs. John D. Rockford invested a certain amount of money at $8\frac{1}{2}\%$, but then invested $600 more at $7\frac{1}{2}\%$. If her total income from both investments is $4421, what amount was invested at $8\frac{1}{2}\%$ and at $7\frac{1}{2}\%$?

9. Mr. Robertson invested $1500 less at 9% than at 8%. If the income from the smaller investment is $95 less than the income from the larger investment, how much did he invest at 9% and at 8%?

10. Three investments yield $1965. The amount invested at 9% is $1500 more than that at 8%. The amount invested at 10% is $500 less than 3 times the amount invested at 8%. How much was invested at 8%, 9%, and 10%?

ANSWERS TO PRACTICE PROBLEMS: INVESTMENT

1. Let x = amount invested at 6%

$1,000 - x$ = amount invested at 7%

Equation: $.06x + .07(1,000 - x) = 65.00$

Move decimal point 2 places to the right

$$6x + 7,000 - 7x = 6,500$$
$$-x = 6,500 - 7,000$$
$$-x = -500$$
$$x = \$500 \text{ invested in 1st bank}$$
$$\$1,000 - x = \$500 \text{ invested in 2nd bank}$$

2. Let x = amount invested in bonds

$12,000 - x$ = amount invested in common stocks

Equation: $.09x + .08(12,000 - x) = 1,040.00$

Move decimal point 2 places to the right

$$9x + 96,000 - 8x = 104,000$$
$$9x - 8x = 104,000 - 96,000$$
$$x = \$8,000 \text{ invested in bonds}$$
$$12,000 - x = \$4,000 \text{ invested in}$$
$$\text{common stocks}$$

3. Let x = amount invested at $8\frac{1}{2}\%$

$14,000 - x$ = amount invested at $7\frac{1}{2}\%$

Equation: $.08\frac{1}{2}(x) + .07\frac{1}{2}(14,000 - x) = 1,152.00$

Move decimal point 2 places to the right

$$8\frac{1}{2}x + 7\frac{1}{2}(14,000 - x) = 115,200$$

$$8\frac{1}{2}x + 105,000 - 7\frac{1}{2}x = 115,200$$

$$x = 115,200 - 105,000$$

$$x = \$10,200$$

$$14,000 - x = \$3,800$$

4. Let x = amount invested at 7%

$2x$ = amount invested at $7\frac{1}{2}\%$

Equation: $.06\frac{1}{2}(2x) + .07(x) = 800.00$

Move decimal point 2 places to the right

$$13x + 7x = 80,000$$

$$20x = 80,000$$

$$x = \$4,000$$

$$2x = \$8,000$$

5. Let x = amount invested at $8\frac{1}{2}\%$ bonds

$18,000 - x$ = amount invested at $7\frac{1}{2}\%$ common stock

Equation: $.08\frac{1}{2}(x) + .07\frac{1}{2}\%(18,000 - x) = \$1,470.00$

Move decimal point 2 places to the right

$$8\frac{1}{2}(x) + 7\frac{1}{2}(18,000 - x) = 147,000$$

$$8\frac{1}{2}x + 135,000 - 7\frac{1}{2}x = 147,000$$

$$x = 147,000 - 135,000$$

$$x = \$12,000$$

$$18,000 - x = \$6,000$$

6. Let x = amount invested at $7\frac{1}{2}\%$

$12,000 - x$ = amount invested at 8%

Equation: $.07\frac{1}{2}x - .08(12,000 - x) = 280.00$

Move decimal point 2 places to the right

$$7\frac{1}{2}x - 96,000 + 8x = 28,000$$

$$15\frac{1}{2}x = 28,000 + 96,000$$

$$15.5x = 124,000$$

$$x = \$8,000$$

$$12,000 - x = \$4,000$$

7.

1st Investment		2nd Investment		Yield
x	$-$	$x-2,000$	$=$	$\$110.00$
$8\frac{1}{2}\%$		$9\frac{1}{2}\%$		

Equation: $.08\frac{1}{2}(x)-.09\frac{1}{2}(x-2,000)=110.00$

Move decimal point 2 places to the right

$$8\frac{1}{2}(x)-9\frac{1}{2}x+19,000=11,000$$

$$8\frac{1}{2}x-9\frac{1}{2}x=11,000-19,000$$

$$-x=-8,000$$

$$x=\$8,000 \text{ invested @ } 8\frac{1}{2}\%$$

$$x-2,000=\$6,000 \text{ invested @ } 9\frac{1}{2}\%$$

8.

1st Investment		2nd Investment		Yield
x	$+$	$x+600$	$=$	$\$4,421.00$
$8\frac{1}{2}\%$		$7\frac{1}{2}\%$		

Equation: $.08\frac{1}{2}(x)+.07\frac{1}{2}(x+600)=4,421.00$

Move decimal point 2 places to the right

$$.08\frac{1}{2}x+7\frac{1}{2}x+4,500=442,100$$

$$16x=442,100-4,500$$

$$16x=437,600$$

$$x=\$27,350 \text{ at } 8\frac{1}{2}\%$$

$$x+600=\$27,950 \text{ at } 7\frac{1}{2}\%$$

9.

1st Investment		2nd Investment	Yield
$\boxed{x}$	$-$	$\boxed{x-1,500}$ $=$	$\boxed{\$95.00}$
8%		9%	

Equation: $.08(x)-.09(x-1,500)=95.00$

Move decimal point 2 places to the right

$$8x-9(x-1,500)=9,500$$
$$8x-9x+13,500=9,500$$
$$-x+13,500=9,500$$
$$-x=9,500-13,500$$
$$x=\$4,000 \text{ at } 8\%$$
$$x-1,500=\$2,500 \text{ at } 9\%$$

10.

1st Investment		2nd Investment		3rd Investment		Yield
$\boxed{x}$	$+$	$\boxed{x+1,500}$	$+$	$\boxed{3x-500}$	$=$	$\boxed{\$1,965.00}$
8%		9%		10%		

Equation: $.08(x)+.09(x+1,500)+.10(3x-500)=1,965.00$

Move decimal point 2 places to the right

$$8x+9x+13,500+30x-5,000=196,500$$
$$47x+8,500=196,500$$
$$47x=196,500-8,500$$
$$47x=188,000$$
$$x=\$4,000 @ 8\%$$
$$x+1,500=\$5,500 @ 9\%$$
$$3x-500=\$11,500 @ 10\%$$

Chapter Eight

Work Problems

Study these complex fractions and their simplification to common fractions. They will improve your ability to solve work problems.

1. $\dfrac{1}{2\frac{1}{2}} = 1 \div 2\frac{1}{2}$

$= 1 \div \dfrac{5}{2}$

$= 1 \cdot \dfrac{2}{5}$

$\dfrac{1}{2\frac{1}{2}} = \dfrac{2}{5}$

3. $\dfrac{2x}{6\frac{2}{3}} = 2x \div 6\frac{2}{3}$

$= 2x \div \dfrac{20}{3}$

$= \cancel{2}x \cdot \dfrac{3}{\cancel{20}\,10}$

$\dfrac{2x}{6\frac{2}{3}} = \dfrac{3x}{10}$

2. $\dfrac{1}{3\frac{1}{3}} = 1 \div 3\frac{1}{3}$

$= 1 \div \dfrac{10}{3}$

$= 1 \cdot \dfrac{3}{10}$

$\dfrac{1}{3\frac{1}{3}} = \dfrac{3}{10}$

4. $\dfrac{4\frac{1}{2}}{7\frac{1}{3}} = 4\frac{1}{2} \div 7\frac{1}{3}$

$= \dfrac{9}{2} \div \dfrac{22}{3}$

$= \dfrac{9}{2} \cdot \dfrac{3}{22}$

$\dfrac{4\frac{1}{2}}{7\frac{1}{3}} = \dfrac{27}{44}$

Before solving work problems, you must be able to solve equations containing fractions. In order to solve these equations, you must have a lowest common denominator (LCD).

EXAMPLE 1:

Solve for x:

Equation: $\dfrac{1}{2} + \dfrac{1}{4} = \dfrac{1}{x}$

Step 1. Find the LCD For this equation, LCD = $4x$.

Step 2. Multiply each numerator of each fraction by the LCD.

Step 3. Divide each numerator by the denominator of each fraction.

Step 4. Solve equation for the value of the letter.

Above equation: $\dfrac{1}{2} + \dfrac{1}{4} = \dfrac{1}{x}$

LCD = $4x$

$$\dfrac{\overset{2}{\cancel{4}x}}{\cancel{2}} + \dfrac{\cancel{4}x}{\cancel{4}} = \dfrac{4\cancel{x}}{\cancel{x}}$$

$$2x + x = 4$$

$$3x = 4$$

$$x = 1\tfrac{1}{3}$$

EXAMPLE 2:

Equation: $\dfrac{15}{2x} + \dfrac{4}{x} = \dfrac{23}{4}$

LCD = $4x$

$$\overset{2}{\cancel{4}}\cancel{x} \cdot \dfrac{15}{\cancel{2}\cancel{x}} + \cancel{4}x \cdot \dfrac{4}{\cancel{x}} = \cancel{4}x\dfrac{23}{\cancel{4}}$$

$$30 + 16 = 23x$$

$$46 = 23x$$

$$2 = x$$

EXAMPLE 3:

Equation: $\dfrac{3x-4}{7} = \dfrac{2x-4}{4}$

LCD = 28 $\dfrac{\overset{4}{\cancel{28}}(3x-4)}{\cancel{7}} = \dfrac{\overset{7}{\cancel{28}}(2x-4)}{\cancel{4}}$

$$12x-16 = 14x-28$$
$$12x-14x = -28+16$$
$$-2x = -12$$
$$x = 6$$

EXAMPLE 4:

Equation: $\dfrac{3}{x+2} + \dfrac{5}{x-2} = \dfrac{28}{(x+2)(x-2)}$

LCD = $(x+2)(x-2)$

$$\cancel{(x+2)}(x-2)\frac{3}{\cancel{x+2}} (x+2)\cancel{(x-2)}\frac{5}{\cancel{x-2}} = \frac{28\cancel{(x+2)}\cancel{(x-2)}}{\cancel{(x+2)}\cancel{(x-2)}}$$
$$3(x-2)+5(x+2) = 28$$
$$3x-6+5x+10 = 28$$
$$3x+5x = 28+6-10$$
$$8x = 24$$
$$x = 3$$

STUDY PROBLEM 1

Bill can mow the lawn in 2 hours, and his younger brother Tom can mow it in 3 hours. If both brothers mow the lawn together, how many hours will it take them?

Model:

Persons	Hours to mow lawn	Part done in 1 hr
Bill	2	$\frac{1}{2}$
Tom	3	$\frac{1}{3}$
Together	x	$\frac{1}{x}$

Equation: $\dfrac{1}{2} + \dfrac{1}{3} = \dfrac{1}{x}$

LCD = $6x$ $3x + 2x = 6$

$5x = 6$

$x = 1\frac{1}{5}$ hr or 1 hr and 12 min

to finish the job together

STUDY PROBLEM 2

One pump can fill a swimming pool in 8 hours and another pump can fill it in 10 hours. If both pumps are opened at the same time, how many hours will it take to fill the pool?

Equipment	Hours to fill pool	Part filled in 1 hr
1st pump	8	$\frac{1}{8}$
2nd pump	10	$\frac{1}{10}$
Together	x	$\frac{1}{x}$

Equation: $\dfrac{1}{8} + \dfrac{1}{10} = \dfrac{1}{x}$ LCD = $40x$ $5x + 4x = 40$

$9x = 40$

$x = 4\frac{4}{9}$ hr

STUDY PROBLEM 3

One pump can fill a reservoir in 60 hours. Another pump can fill the same reservoir in 80 hours. A third pump can empty the reservoir in 90 hours. If all three pumps are operating at the same time, how long will it take to fill the reservoir?

Equipment	Hours to fill reservoir	Part filled in 1 hr
1st pump	60	$\frac{1}{60}$
2nd pump	80	$\frac{1}{80}$
3rd pump	90	$\frac{1}{90}$
Together	x	$\frac{1}{x}$

Equation: $\dfrac{1}{60}+\dfrac{1}{80}-\dfrac{1}{90}=\dfrac{1}{x}$ $LCD = 720x$ $12x + 9x - 8x = 720$

$$13x = 720$$

$$x = 55\tfrac{5}{13}\ hr$$

STUDY PROBLEM 4

One printing press can print 5,000 advertising cards in 12 seconds. Another printing press can print the same number of cards in 7½ seconds. If both presses are used together to print the 5,000 cards, how many seconds will it take them?

Equipment	Seconds to print job	Part done in 1 sec
1st press	12	$\frac{1}{12}$
2nd press	$7\frac{1}{2}$	$\frac{1}{7\frac{1}{2}}$
Together	x	$\frac{1}{x}$

Equation: $\dfrac{1}{12}+\dfrac{1}{7\frac{1}{2}}=\dfrac{1}{x}$ $LCD = 60x$

Multiply each numerator by the LCD

Divide each numerator by each denominator.

Solve equation.

$$\frac{1}{12}+\frac{2}{15}=\frac{1}{x}$$

$$^{5}\cancel{60}\,x\,\frac{(1)}{\cancel{12}}+\,^{4}\cancel{60}\,x\,\frac{(2)}{\cancel{15}}=60\,\cancel{x}\,\frac{1}{\cancel{x}}$$

$$5x+8x=60$$

$$13x=60$$

$$x=4\,^{8}\!/_{13}\text{ sec}$$

STUDY PROBLEM 5

Willard can paint a living room in 6 hours. His assistant can paint the same room in 8 hours. If Willard paints for 1 hour and then leaves, letting his assistant take over, how long will it take the assistant to complete the job?

Note: Problems that involve 1 or more workers who perform at various rates of speed and are then replaced by other workers can be solved with the following format.

ABCD is a square that represents the whole job. The shaded area represents $^{1}\!/_{6}$ of the job. The blank area, $^{5}\!/_{6}$ of the job.

Persons	Hours to complete job	Part done in 1 hr
Willard	6	$\frac{1}{6}$
Assistant	8	$\frac{1}{8}$
Together	x	$\frac{1}{x}$
Part to be completed	----------------------------------- $\frac{5}{6}$ of job	

Since Willard left the job, he is no longer in the equation.

Equation: $\frac{1}{8} = \frac{1}{x} \cdot \frac{5}{6}$ Part to be completed by assistant

LCD = 24x $^3\cancel{24}x\left(\frac{1}{\cancel{8}}\right) = {}^4\cancel{24}\cancel{x}\left(\frac{5}{6\cancel{x}}\right)$

$$3x = 20$$

$$x = 6\frac{2}{3} \text{ hr to complete the job}$$

STUDY PROBLEM 6

One computer can do a printout for a job in 7 hours. Another computer can do the same printout in 8 hours. If the first computer is used for 2 hours and then the second computer is used, how long will it take the second computer to complete the job?

Equipment	Hours to complete job	Part done in 1 hr
1st computer	7	$\frac{1}{7}$
2nd computer	8	$\frac{1}{8}$
Together	x	$\frac{1}{x}$
Part to be completed	----------------------------------- $\frac{5}{7}$ of job	

Equation: $\frac{1}{8} = \frac{1}{x} \cdot \frac{5}{7} \rightarrow$ Part to be completed

$$\text{LCD} = 56x \qquad\qquad 7x = 40$$
$$x = 5\tfrac{5}{7} \text{ hr for 2nd computer}$$

STUDY PROBLEM 7

Bess can type an attorney's brief in 6 hours. Betty can type the same brief in 8 hours. If Bess works 1 hour and stops, and then Betty continues to complete the job, how long will it take Betty to complete the job alone?

Persons	Hours to complete job	Part done in 1 hr
Bess	6	$\frac{1}{6}$
Betty	8	$\frac{1}{8}$
Together	x	$\frac{1}{x}$
Part to be completed	------------------------------------	$\frac{5}{6}$ of job

Equation: $\quad \dfrac{1}{8} = \dfrac{1}{x} \cdot \dfrac{5}{6} \rightarrow$ Part to be completed

$$\dfrac{1}{8} = \dfrac{5}{6x}$$

$$\text{LCD} = 24x \qquad\qquad 3x = 20$$
$$x = 6\tfrac{2}{3} \text{ hr for Betty to complete the job}$$

PRACTICE PROBLEMS: WORK

1. It takes Mr. Rose 10 hours to complete a printing job. His helper John can complete the same job in 15 hours. If Mr. Rose and John work together, how long will it take them to complete the job?

2. It takes one machine 25 minutes to pick a bale of cotton and another machine 30 minutes. If both machines are used, how long will it take them to pick a bale of cotton?

3. A tank can be filled by one pump in 50 minutes and by another pump in 60 minutes. A third pump can drain the tank in 75 minutes. If all 3 pumps go into operation, how long will it take to fill the tank?

4. Nancy can do a typing job in 8 hours. When Carole helps her, they can do the job together in 5 hours. How many hours would it take Carole to do the job alone?

5. Mr. Sharrone can paper a room in 3 hours. His assistant needs 5 hours. If Mr. Sharrone works 1 hour and then his assistant completes the job alone, how long will it take the assistant to finish the job?

6. Mr. Rodriguez, a college instructor, can grade his class papers in 3 hours while it takes his assistant 4½ hours. If Mr. Rodriguez graded the papers for 1 hour and then left the job for his assistant to complete, how long will it take his assistant to finish grading papers?

7. Mr. Ito's assistant takes twice as long to complete a computer task as Mr. Ito. If it takes both experts 6 hours to complete the task, how long will it take each of them to do the job alone?

8. Mr. Vella can build a brick wall in 4 days. His apprentice can build the same wall in 6 days. After working alone for 3 days, Mr. Vella became ill and left the job for his apprentice to complete. How many days did it take the apprentice to finish the wall?

ANSWERS TO PRACTICE PROBLEMS: WORK

1.

Persons	Hours to complete job	Part done in 1 hr
Mr. Rose	10	$\frac{1}{10}$
John	15	$\frac{1}{15}$
Together	x	$\frac{1}{x}$

Equation: $\dfrac{1}{10}+\dfrac{1}{15}=\dfrac{1}{x}$ LCD $=30x$ $3x+2x=30$

$5x=30$

$x=6$ hr to complete job together

2.

Equipment	Minutes to pick bale	Part done in 1 hr
1st machine	25	$\frac{1}{25}$
2nd machine	30	$\frac{1}{30}$
Together	x	$\frac{1}{x}$

Equation: $\dfrac{1}{25} + \dfrac{1}{30} = \dfrac{1}{x}$

LCD = 150x $\quad 6x + 5x = 150$

$$11x = 150$$

$$x = 13\frac{7}{11} \text{ min to pick a bale}$$

3.

Equipment	Minutes to fill tank	Part filled in 1 min
1st pump	50	$\frac{1}{50}$
2nd pump	60	$\frac{1}{60}$
3rd pump	75 drain (−)	$-\frac{1}{75}$
Together	x	$\frac{1}{x}$

Equation: $\dfrac{1}{50} + \dfrac{1}{60} - \dfrac{1}{75} = \dfrac{1}{x}$

LCD = 300x $\quad 6x + 5x - 4x = 300$

$$7x = 300$$

$$x = 42\frac{6}{7} \text{ min to fill tank}$$

4.

Persons	Hours to complete job	Part done in 1 hr
Nancy	8	$\frac{1}{8}$
Carole	x	$\frac{1}{x}$
Together	5	$\frac{1}{5}$

Equation: $\dfrac{1}{8}+\dfrac{1}{x}=\dfrac{1}{5}$

LCD = 40x $5x + 40 = 8x$

$40 = 8x - 5x$

$40 = 3x$

$13\frac{1}{3} = x$ hr for Carole to do the job alone

5.

Persons	Hours to complete job	Part done in 1 hr
Sharrone	3	$\frac{1}{3}$
Assistant	5	$\frac{1}{5}$
Together	x	$\frac{1}{x}$

Part to be completed ----------------------- $\frac{2}{3}$ of job

Equation: $\dfrac{1}{5}=\dfrac{1}{x}\cdot\dfrac{2}{3}$

LCD = 15x $\dfrac{1}{5}=\dfrac{2}{3x}$

$3x = 10$

$x = 3\frac{1}{3}$ hr for assistant to complete job

6.

Persons	Hours to complete job	Part done in 1 hr
Mr. Rodriguez3	$\frac{1}{3}$	
Assistant	$4\frac{1}{2}$	$\frac{2}{9}$
Together	x	$\frac{1}{x}$

Part to be completed ------------------------------------- $\frac{2}{3}$ of job

Equation: $\dfrac{2}{9} = \dfrac{1}{x} \cdot \dfrac{2}{3}$

LCD = 9x $\qquad \dfrac{2}{9} = \dfrac{2}{3x}$

$\qquad\qquad\qquad 2x = 9$

$\qquad\qquad\qquad x = 3$ hr for assistant to finish grading papers

7.

Persons	Hours to complete job	Part done in 1 hr
Mr. Ito	x	$\frac{1}{x}$
Assistant	$2x$	$\frac{1}{2x}$
Together	6	$\frac{1}{6}$

Part to be completed ------------------------------------- whole job

Equation: $\dfrac{1}{x} + \dfrac{1}{2x} = \dfrac{1}{6}$

LCD = 6x $\qquad 6 + 3 = x$

$\qquad\qquad\qquad x = 9$ hr for Mr. Ito

$\qquad\qquad\qquad 2x = 18$ hr for assistant

8.

Persons	Days to do job	Part done in 1 day
Mr. Vella	4	$\frac{1}{4}$
Apprentice	6	$\frac{1}{6}$
Together	x	$\frac{1}{x}$

Part to be completed ------------------------------------ $\frac{1}{4}$ of job

Equation: $\dfrac{1}{6} = \dfrac{1}{x} \cdot \dfrac{1}{4}$

LCD $= 12x$ $^{2}\cancel{12}x\left(\dfrac{1}{\cancel{6}}\right)=\left(\dfrac{1}{\cancel{4}\,\cancel{x}}\right)^{3}\cancel{12}\,\cancel{x}$

$ 2x = 3$

$ x = 1\frac{1}{2}$ days for apprentice to complete the job

Chapter Nine

Attendance Problems

To solve problems that involve attendance at events, you can use the same model used in solving investment problems.

Model:

Number of adult tickets	+	Number of children's tickets	=	Total amount of ticket receipts
Cost per ticket		Cost per ticket		

STUDY PROBLEM 1

The admission at a concert was $7.50 for adults and $3 for children. If the total gate receipts were $4050 and there were 600 people in attendance, how many adult tickets and how many children's tickets were sold?

Adult tickets	+	Children's tickets	=	Receipts
x		$(600-x)$		$4,050.00
$7.50		$3.00		

  a. Move decimal point 2 places to the right.

  b. Multiply cost of each ticket by the number of tickets.

Equation: $7.50(x) + 3.00(600-x) = 4,050.00$

$$750x + 300(600-x) = 405,000$$
$$750x + 180,000 - 300x = 405,000$$
$$450x = 405,000 - 180,000$$
$$450x = 225,000$$
$$x = 500$$
$$600 - x = 100$$

PRACTICE PROBLEMS: ATTENDANCE

1. The admission at a local football game was $3.50 for adults and $1.25 for students and children. The receipts were $4575 for $1500 admissions. How many of each kind of ticket were sold?

2. On opening night, 1,500 tickets were sold for a college play. Adult tickets were sold at $5.25 each and student tickets at $2.25 each. If the total receipts collected were $4875, how many of each kind of ticket were sold?

3. At a recent ballet school performance, the number of young adult tickets sold was 50 less than twice the number of adult tickets sold, and the number of children's tickets sold was the same as the number of adult tickets. If adult tickets cost $2.50 each, young adult tickets $1.50 each, and children's tickets $1 each, how many of each kind of ticket were sold if the cash receipts amounted to $705?

4. At a university football game, 5,000 more regular-seat tickets were sold at $7.50 each than box office seats at $12.50 each. If the receipts collected amounted to $637,000, how many tickets were sold at $7.50 each and how many were sold at $12.50 each?

See answers that follow.

ANSWERS TO PRACTICE PROBLEMS: ATTENDANCE

1.

Adult tickets	+	Student and children's tickets	=	Total receipts
x		$1,500 - x$		$4,575.00
$3.50		$1.25		

Equation:

$$3.\underline{50}(x) + 1.\underline{25}(1,500 - x) = 4,575.\underline{00}$$

Move decimal point 2 places to the right

$$350x + 125(1,500 - x) = 457,500$$

$$350x + 187,500 - 125x = 457,500$$

$$350x - 125x = 457,500 - 187,500$$

$$225x = 270,000$$

$$x = 1,200 \text{ adult tickets sold}$$

$$1,500 - x = 300 \text{ student and children's tickets sold}$$

2.

Adult tickets	+	Student tickets	=	Total receipts
x		$1,500 - x$		$4,875.00
$5.25		$2.25		

Equation: $5.\underline{25}(x) + 2.\underline{25}(1,500 - x) = \$4,875.\underline{00}$

Move decimal point 2 places to the right

$$525x + 225(1,500 - x) = 487,500$$

$$525x + 337,500 - 225x = 487,500$$

$$300x = 487,500 - 337,500$$

$$300x = 150,000$$

$$x = 500 \text{ adult tickets}$$

$$1,500 - x = 1,000 \text{ student tickets}$$

3.

Adult tickets		Young Adult tickets		Children's tickets		Total receipts
x	+	$2x - 50$	+	x	=	$705.00
$2.50		$1.50		$1.00		

Equation:

$$2.50(x) + 1.50(2x - 50) + 1.00(x) = \$705.00$$

Move decimal point 2 places to the right

$$250x + 150(2x - 50) + 100x = 70,500$$
$$250x + 300x - 7,500 + 100x = 70,500$$
$$650x - 7,500 = 70,500$$
$$650x = 70,500 + 7,500$$
$$650x = 78,000$$
$$x = 120 \text{ adult tickets}$$
$$x = 120 \text{ children's tickets}$$
$$2x - 50 = 190 \text{ young adult tickets}$$

4.

Box office-seat tickets		Regular-seat tickets		Total receipts
x	+	$5,000 + x$	=	$637,500.00
$12.50		$7.50		

Equation:

Chapter Ten

Lever Problems

A diagram is especially helpful when working with lever problems.

A lever is a stiff board or bar which is balanced upon a support called a fulcrum. AC and CB are called lever arms, and C is the fulcrum. If weight A and weight B are equal, AC will balance CB if C, the fulcrum, is in the middle. If you played on a teeterboard or seesaw when you were a child, you know that two people will balance only when the heavier of the two sits closer to the fulcrum.

FORMULA: $W(A) \cdot d1 = d2 \cdot W(B)$

STUDY PROBLEM 1

A 40-pound weight placed 2 feet from the fulcrum of a lever balances an unknown weight placed 4 feet from the fulcrum. Find the unknown weight.

Diagram:

Equation:
$$40 \cdot 2 = 4 \cdot x$$
$$80 = 4x$$
$$x = 20 \text{ lb}$$

PRACTICE PROBLEMS: LEVERS

1. A weight of 105 pounds is 4 feet from the fulcrum of a lever, and an unknown weight is 6 feet from the fulcrum. What is the unknown weight if the lever is balanced?

2. A 90-pound weight is placed on a lever at 7 feet from the fulcrum. At what distance from the fulcrum must a 63-pound weight be placed to make the lever balance?

3. A 3,000-pound weight rests on one end of a lever and a 300-pound weight rests on the other end. If the lever is 22 feet long, where must the fulcrum be placed to make the lever balance?

4. A heavy boulder had to be removed from a plot of land so that the builder could begin excavating for the foundation of a new home. The builder decided to use the lever principle to move the boulder. If the boulder is 1 foot from the fulcrum and a 300-pound weight is 3 feet from the fulcrum, what is the weight of the boulder? (Draw a diagram.)

ANSWERS TO PRACTICE PROBLEMS: LEVERS

1. Equation: $105 \cdot 4 = 6 \cdot x$

$420 = 6x$

$70 \text{ lb} = x$

2. Equation: $90 \cdot 7 = x \cdot 63$

$630 = 63x$

$10 \text{ ft} = x$

3. Equation: $3,000 \cdot x = 300(22 - x)$

$3,000x = 6,600 - 300x$

$3,300x = 6,600$

$x = 2$ ft from the 3,000-lb weight

4.

$$W \bullet 1 = 3 \bullet (300)$$
$$W1 = 900 \text{ lb weight of boulder}$$

Chapter Eleven

Age Problems

It's important to understand the translation of English into algebraic symbols. For example:

I.   a. Four years from now = +4

   b. Four years ago = –4

   c. Sherrie is 3 times as old as her sister Susie. Five years from now, Sherrie will be twice as old as Susie. How old is each now?

Fill in the question marks:

Model:

Persons	Age now	Age 5 years from now
Sherrie	?	?
Susie	x	?

Equation: ? = 2(?)

II. If Bill's age now is x years, represent the following algebraically:

   a. Bill's age 4 years from now

   b. Bill's age 6 years ago

   c. $\frac{2}{3}$ of Bill's age 1 year ago

   d. $\frac{3}{4}$ of Bill's age 2 years from now

III. The sum of Anne's age and her mother's age is 42 years. If x is Anne's age, her mother's age is $42 - x$ years.

ANSWERS TO AGE PROBLEMS I AND II

I.

Persons	Age now	Age 5 years from now
Sherrie	$3x$	$3x + 5$
Susie	x	$x + 5$

Equation: $3x + 5 = 2(x + 5)$

$$3x + 5 = 2x + 10$$
$$x = 5 \ (\text{Susie's age})$$
$$3x = 15 \ (\text{Sherrie's age})$$

II. a. $x + 4$

b. $x - 6$

c. $\frac{2}{3}(x - 1)$

d. $\frac{3}{4}(x + 2)$

STUDY PROBLEM 1

Juan is 5 years older than his brother Carlos. Five years ago, Juan was twice as old as Carlos. How old is each now?

Persons	Age now	Age 5 years ago
Juan	$x + 5$	$x + 5 - 5$
Carlos	x	$x - 5$

Equation: $x + 5 - 5 = 2(x - 5)$

$$x = 2(x - 5)$$
$$x = 2x - 10$$
$$10 = 2x - x$$
$$10 = x \ (\text{Carlos's age now})$$
$$15 = x + 5 \ (\text{Juan's age now})$$

STUDY PROBLEM 2

The sum of John's age and his brother Jerry's age is 42. Eight years ago John's age was 2 years more than Jerry's. How old is each now?

Persons	Age now	Age 8 years ago
John	$42 - x$	$42 - x - 8$
Jerry	x	$x - 8$

Equation: $42 - x - 8 = x - 8 + 2$

$34 - x = x - 6$

$-x - x = -6 - 34$

$-2x = -40$

$x = 20 \ (\text{Jerry's age})$

$42 - x = 22 \ (\text{John's age})$

STUDY PROBLEM 3

Kathryn is 8 years older than her sister Rosemary. Eight years ago Kathryn was 3 times as old as Rosemary. How old is each now?

Model:

Persons	Age now	Age 8 years ago
Kathryn	$x + 8$	$x + 8 - 8$
Rosemary	x	$x - 8$

Equation: $x = 3(x - 8)$

$x = 3x - 24$

$x - 3x = -24$

$-2x = -24$

$x = 12 \ (\text{Rosemary's age now})$

$x + 8 = 20 \ (\text{Kathryn's age now})$

PRACTICE PROBLEMS: AGE

1. Mr. Kim is 25 years older than his daughter. In 10 years Mr. Kim will be twice as old as his daughter will be. How old is each now?

2. Mr. Polaski is 5 times as old as his son David. Five years ago the father was 15 times as old as his son. What are their ages now?

3. The sum of Michael's age and his pal Ronnie's age is 41. Six times Michael's age 10 years ago was equal to 2 more than twice Ronnie's age now. How old is each now?

4. Mrs. Tracey is now 3 times as old as her daughter Juliana. Five years from now, Mrs. Tracey will be 7 times her daughter's age 5 years ago. What are the mother's and daughter's ages now?

5. If Tina's age 1 year from now plus her age 2 years ago plus three times her age 5 years ago is equal to 69 years, how old is she now?

Check your answers with the solutions that follow.

ANSWERS TO PRACTICE PROBLEMS: AGE

1.

Persons	Age now	Age in 10 years
Mr. Kim	$x + 25$	$x + 35$
Daughter	x	$x + 10$

Equation:

$$x + 35 = 2(x + 10)$$
$$x + 35 = 2x + 20$$
$$-2x + x = -35 + 20$$
$$-x = -15$$
$$x = 15 \ (\text{daughter's age})$$
$$x + 25 = 40 \ (\text{Mr. Kim's age})$$

2.

Persons	Age now	Age 5 years ago
Mr. Polaski	$5x$	$5x - 5$
David	x	$x - 5$

Equation:

$$5x - 5 = 15(x - 5)$$
$$5x - 5 = 15x - 75$$
$$5x - 15x = -70$$
$$-10x = -70$$
$$x = 7 \ (\text{David's age})$$
$$5x = 35 \ (\text{Mr. Polaski's age})$$

3.

Persons	Age now	Age 8 years ago	Age 5 years ago
Carole	$5x + 8$	$5x$	$5x + 3$
Carl	$x + 8$	x	$x + 3$

Equation:
$$5x + 3 = 3(x + 3)$$
$$5x + 3 = 3x + 9$$
$$5x - 3x = 9 - 3$$
$$2x = 6$$
$$x = 3 \text{ (Carl's age 8 years ago)}$$
$$x + 8 = 11 \text{ (Carl's age now)}$$
$$5x + 8 = 23 \text{ (Carole's age now)}$$

4.

Persons	Age now	Age 5 yr from now	Age 5 yr ago
Mrs. Tracey	$3x$	$3x + 5$	?
Juliana	x	?	$x - 5$

Equation:
$$3x + 5 = 7(x - 5)$$
$$3x + 5 = 7x - 35$$
$$3x - 7x = -35 - 5$$
$$-4x = -40$$
$$x = 10 \text{ (Juliana's age now)}$$
$$3x = 30 \text{ (Mrs. Tracey's age now)}$$

5.

Tina's age now	1 yr from now	2 yr from now	5 yr ago
x	$x + 1$	$x - 2$	$x - 5$

Equation:
$$(x + 1) + (x - 2) + 3(x - 5) = 69$$
$$x + 1 + x - 2 + 3x - 15 = 69$$
$$5x - 16 = 69$$
$$5x = 69 + 16$$
$$5x = 85$$
$$x = 17 \text{ (Tina's age now)}$$

Chapter Twelve

Ratio Problems

Ratio is the relationship in quantity, amount or size between two or more things. We can compare numbers rather easily. For example, if John has 10 dollars and Bill has 5 dollars, then we can say that John has 5 dollars more than Bill or we can say that he has twice as much as Bill. We can express this comparison or ratio in a variety of ways:

$$\frac{2}{1}$$

$$2 : 1$$

$$2 \text{ to } 1$$

$$2 \div 1$$

When we compare two rational numbers by division, we have a ratio. (A rational number is the quotient of any two integers* divided by any number except zero.)

EXAMPLES:

a. $\dfrac{2}{5}$

b. $\dfrac{13}{5}$

c. $\dfrac{126}{17}$

d. $\dfrac{.03}{6.25}$

e. $\dfrac{-12.50}{-6.25}$

f. $\dfrac{+412 \text{ cm}}{-4 \text{ cm}}$

g. $\dfrac{-18 \text{ gal}}{+2.5 \text{ gal}}$

h. $\dfrac{-12\frac{1}{2} \text{ yd}}{-.6 \text{ yd}}$

i. $\dfrac{+16\frac{2}{3}}{-6\frac{2}{3}}$

All ratios must be reduced to their lowest terms and have the same unit of measure.

EXAMPLES:

a. $\dfrac{12}{18} = \dfrac{2}{3}$

b. $\dfrac{25}{60} = \dfrac{5}{12}$

c. $\dfrac{-4 \text{ ft}}{+30 \text{ in.}} = \dfrac{-48 \text{ in.}}{30 \text{ in.}} = \dfrac{-8}{+5}$

d. $\dfrac{-2 \text{ dollars}}{-60 \text{ cents}} = \dfrac{200 \text{ cents}}{60 \text{ cents}} = \dfrac{+10}{+3}$

e. $\dfrac{2 \text{ yd}}{1\!/_2 \text{ yd}} = 2 \div \dfrac{1}{2} = 2 \cdot \dfrac{2}{1} = \dfrac{4}{1}$

f. $\dfrac{-150 \text{ cm}}{-2 \text{ meters}} = \dfrac{150 \text{ cm}}{200 \text{ cm}} = \dfrac{+3}{+4}$

* An integer is any number that is a member of a set of positive and negative whole numbers including zero.

g. $\dfrac{4\,^1\!/_2 \text{ gal}}{4 \text{ qt}} = \dfrac{18 \text{ qt}}{4 \text{ qt}} = \dfrac{9}{2}$

h. $\dfrac{25 \text{ min}}{1 \text{ hr}} = \dfrac{25 \text{ min}}{60 \text{ min}} = \dfrac{5}{12}$

i. $\dfrac{3x \text{ ft}}{5x \text{ ft}} = \dfrac{3}{5}$

PRACTICE PROBLEMS: RATIOS

Write the following ratios and reduce to lowest terms.

1. $\dfrac{32}{18}$

2. 27 : 81

3. 330 to 99

4. 16 ft to 4 yds

5. 38 cm : 57 cm

6. 300 m to 1 km

7. 108 in. : 3 ft

8. Golf balls formerly sold at $1 each. Now on sale, they sell for $9 a dozen. What is the ratio of the former price to the sale price?

9. Express as a ratio the number of 34-cent stamps you can buy if you have (C) cents.

10.

10 in.

18 in.

What is the ratio of the perimeter of square (a) to the perimeter of square (b)?

11.

What is the ratio of the perimeter of rectangle (a) to the perimeter of rectangle (b)?

12.

What is the ratio of the perimeter of triangle (b) to the perimeter of triangle (a)?

13.

What is the ratio of the perimeter of hexagon (a) to the perimeter of hexagon (b)?

14. The budget of a small city is $15,000,000 for the year while yearly the budget for the state is $70,000,000. What is the ratio of the budget for the city to that for the state?

ANSWERS TO PRACTICE PROBLEMS: RATIOS

1. $\dfrac{16}{9}$

2. $\dfrac{1}{3}$

3. $\dfrac{10}{3}$

4. $\dfrac{4}{3}$

5. $\dfrac{2}{3}$

6. $\dfrac{3}{10}$

7. $\dfrac{3}{1}$

8. $\dfrac{4}{3}$

9. $\dfrac{\text{(C) cents}}{34}$

10. $\dfrac{5}{9}$

11. $\dfrac{16}{29}$

12. $\dfrac{21}{13}$

13. $\dfrac{41}{47}$

14. $\dfrac{3}{14}$

Chapter Thirteen

Proportion

When one ratio equals another ratio, you have a proportion.

EXAMPLES:

a. $\dfrac{10}{5} = \dfrac{2}{1}$

e. $\dfrac{6.5}{.5} = \dfrac{13}{1}$

b. $\dfrac{9}{12} = \dfrac{3}{4}$

f. $\dfrac{4\frac{1}{2}}{9} = \dfrac{1}{2}$

c. $\dfrac{27}{6} = \dfrac{9}{2}$

g. $\dfrac{135}{50} = \dfrac{27}{10}$

d. $\dfrac{5}{2\frac{1}{2}} = \dfrac{2}{1}$

h. $\dfrac{.050}{.25} = \dfrac{1}{5}$

Each proportion has 4 terms: numerators and denominators are called terms.

$$\begin{array}{c}\text{Term }(1)\\\text{Term }(2)\end{array}\ \dfrac{8}{12} = \dfrac{2}{3}\ \begin{array}{c}\text{Term }(3)\\\text{Term }(4)\end{array}$$

Terms (1) and (4) are called the *extremes*.

Terms (2) and (3) are called the *means*.

In a proportion, the product of the means is equal to the product of the extremes.

EXAMPLE:

$$\text{Proportion}\ \dfrac{9}{15} = \dfrac{3}{5}$$

$$15 \times 3 \text{ (the means)} = 9 \times 5 \text{ (the extremes)}$$

$$\text{product } 45 = \text{product } 45$$

STUDY PROBLEM 1

Arnold drives 140 miles in 6 hours. Driving at the same rate, how long will it take him to make a trip of 210 miles?

Let x = number of hours to make a trip of 210 miles

$$\text{Proportion: } \frac{140 \text{ miles}}{6 \text{ hours}} = \frac{210 \text{ miles}}{x \text{ hours}}$$

Using the law that the product of the means equals the product of the extremes, we have:

$$140(x) = 6(210)$$
$$140x = 1,260$$
$$x = 9 \text{ hours}$$

STUDY PROBLEM 2

A telephone pole 60 feet high casts a shadow 80 feet long at the same time that a tree casts a shadow 120 feet. What is the height of the tree?

Let x = height of the tree

$$\text{Proportion: } \frac{60 \text{ ft height of pole}}{x \quad \text{height of tree}} = \frac{80 \text{ ft shadow of pole}}{120 \quad \text{shadow of tree}}$$

60 ft height

Shadow is 80 ft

x ft

Tree

Shadow is 120 ft

Law: Product of means equals product of extremes

$$80 \cdot x = 60 \cdot 120$$
$$80x = 7,200$$
$$x = 90 \text{ ft (height of tree)}$$

STUDY PROBLEM 3

If 18 gallons of gasoline cost $23.31, how much would 12 gallons cost?

Let x = cost for 12 gallons

$$\text{Proportion:} \frac{18 \ \text{gal}}{23.31 \ \text{cost}} = \frac{12 \ \text{gal}}{x \ \text{cost}}$$

$$18(x) = 12(23.31)$$

$$18x = \$279.72$$

$$x = \$15.54$$

PRACTICE PROBLEMS: PROPORTION

1. If 12 pencils cost 42 cents, how much would 100 pencils cost? Substitute your solutions for the question marks.

Let x = __?__

Proportion: $\dfrac{?}{?} = \dfrac{?}{?}$

Answer: __?__

2. It takes a crew of 6 men to wash 84 windows of a hospital in one day. At the same rate, how many men would it take to wash 140 windows? Substitute your solution for the question marks.

Let x = __?__

Proportion: $\dfrac{?}{?} = \dfrac{?}{?}$

Answer: __?__

3. If it takes 10.5 bushels of grapes to make 30 quarts of wine, how many bushels of grapes will it take to make 3,200 quarts of wine?

Let $x = $ _?_

Proportion: $\dfrac{?}{?} = \dfrac{?}{?}$

Answer: _?_

4. A hotel 300 feet high casts a 240-foot shadow at the same time that a taller hotel casts a 360-foot shadow. How high is the taller hotel?

Let $x = $ height of the taller hotel

Proportion: $\dfrac{?}{?} = \dfrac{?}{?}$

Answer: _?_

5. A woman died and left an estate valued at $220,000 to her grandson and to her son. If the ratio of the grandson's inheritance to the son's inheritance is 3 : 8, how much money did the son and grandson receive?

Let $3x = $ grandson's inheritance

Let $8x = $ son's inheritance

The sum of the two inheritances = _?_

Equation is _?_

Grandson received _?_

Son received _?_

ANSWERS TO PRACTICE PROBLEMS: PROPORTION

1. x = cost of 100 pencils

Proportion: $\dfrac{12}{.42} = \dfrac{100}{x}$

$$12(x) = .42(100)$$
$$12x = 42.00$$
$$x = \$3.50 \text{ cost of 100 pencils}$$

2. Let x = number of men to wash 140 windows

Proportion: $\dfrac{6 \text{ men}}{84 \text{ windows}} = \dfrac{x \text{ men}}{140 \text{ windows}}$

$$84(x) = 6(140)$$
$$84x = 840$$
$$x = 10 \text{ men}$$

3. Let x = bushels of grapes to make 3,200 quarts of wine

Proportion: $\dfrac{10.5 \text{ bushels}}{30 \text{ quarts}} = \dfrac{x \text{ bushels}}{3,200 \text{ quarts}}$

$$10.5(3,200) = 30(x)$$
$$33,600 = 30x$$
$$1,120 = x \text{ bushels}$$

4.

Proportion: $\dfrac{300 \text{ ft}}{240} = \dfrac{x}{360}$

$$240x = 300(360)$$
$$240x = 108,000$$
$$x = 450 \text{ ft height of taller hotel}$$

5. $11x =$ sum of the two inheritances

Equation: $11x = \$220,000$
$$x = \$\ 20,000 \text{ one share}$$
$$3x = \$\ 60,000 \text{ grandson's share}$$
$$8x = \$160,000 \text{ son's share}$$

Chapter Fourteen

Inequalities

Symbols:

a.   > is greater than

b.   < is less than

c.   = is equal to

d.   ≥ is greater than or is equal to

e.   ≤ is less than or is equal to

So far, we have solved equations where both sides of the equation were equal. In the equation $6x = 24$, for example, x is *equal* to 4.

Sometimes, however, both sides of an equation are *not equal*. In these cases, the solution to the equation may be a series or set of integers greater than or less than a given number.

EXAMPLE 1:

$$x + 3 < 7$$

In this example, x is any positive integer which, when added to 3, is *less* than 7. The solution, then, is not one number but a set of whole numbers.

The *solution set* for the equation above can be described by dots on a graph as follows.

$$\underline{\quad} \cdot \underline{\quad} \cdot \underline{\quad} \cdot \underline{\quad} \cdot \underline{\quad}$$

$$0 \quad 1 \quad 2 \quad 3 \quad 4 \quad 5$$

Keep in mind that zero (0) is not a positive integer, therefore a dot is omitted.

The above graph represents the solution set of $x + 3 < 7$.   Solution Set: [1, 2, 3]

EXAMPLE 2:

$x + 3 > 7$ Solution Set : $[5, 6, 7, ... \textit{ad infinitum}]$

The above graph indicates that 4 is not a member of the solution set, but every point greater than 4 is a member of the solution set.

EXAMPLE 3:

$$-1 \le x \le 4$$

The solution set for the inequality above is in between −1 and +4.

EXAMPLE 4:

a. If $x > 4$, the solution set is [5, 6, 7, 8, ... *ad infinitum*]. If we add 5 to each side of the inequality we have $x + 5 > 4 + 5$.

b. Then $x + 5 > 9$ [5, 6, 7, 8, ...]

Notice (a) and (b) have the same solution set, and this suggests that we have axioms for inequalities.

AXIOM 1

If the same number is added to both members of the inequality, the result is an equivalent inequality.

EXAMPLE 5:

a. If $x > 4$ Solution Set: [5, 6, 7, 8, ...]

Subtracting 2 from each side, we have $x - 2 > 4 - 2$.

b. Then, $x - 2 > 2$ Solution Set: [5, 6, 7, 8, ...]

Notice (a) and (b) have the same solution set.

AXIOM 2

If the same number is subtracted from both members of the inequality, the result is an equivalent inequality.

EXAMPLE 6:

a.   If $x > 4$ Solution Set: [5, 6, 7, 8, ...]

Multiplying both sides of the inequality by 3, we have:

b. $3x > 12$ Solution Set: [5, 6, 7, 8, ...]

Notice (a) and (b) have the same solution set.

AXIOM 3

If both members of an inequality are multiplied by the same number, the result is an equivalent inequality.

EXAMPLE 7:

a.   If $x > 4$ Solution Set: [5, 6, 7, 8, ...]

Dividing both sides of the inequality by 2, we have

$$\frac{x}{2} > \frac{4}{2}.$$

b.   Then $\frac{x}{2} > 2$ Solution Set: [5, 6, 7, 8, ...]

Notice (a) and (b) have the same solution set.

AXIOM 4

If the same number is divided into each member of the inequality, the result is an equivalent inequality.

EXAMPLE 8:

a. If $x > 4$ Solution Set: [5, 6, 7, 8, ...]

Multiplying both sides by –3 we have:

b. $-3x > -12$ Solution Set: [0, 1, 2, 3]

BUT if you reverse the inequality to:

c. $-3x < -12$ Solution Set: [5, 6, 7, 8, ...]

Notice (a) and (c) are equal now.

AXIOM 5

If both members of an inequality are multiplied by a negative number, the inequality sign must be reversed.

EXAMPLE 9:

a. If $x > 4$ Solution Set: [5, 6, 7, 8, ...]

Dividing both sides by – 2 we have

b. $\dfrac{x}{-2} > \dfrac{4}{-2}$ Solution Set: [0, 1, 2]

c. $\dfrac{x}{-2} < -2$ Solution Set: [5, 6, 7, 8, ...]

Notice (a) and (c) have the same solution set.

AXIOM 6

If both members of an inequality are divided by a negative number, the sum of the inequality is reversed.

Study the solutions provided for the inequalities that follow:

STUDY PROBLEM 1

$$4t + 3 > 3t - 7$$
$$4t - 3t > -7 - 3$$
$$t > -10$$

STUDY PROBLEM 2

$$2x - 5 < -x + 10$$
$$2x + x < +10 + 5$$
$$3x < 15$$
$$x < 5$$

STUDY PROBLEM 3

$$3(x + 2) > -2(x + 2)$$
$$3x + 6 > -2x - 4$$
$$3x + 2x > -4 - 6$$
$$5x > -10$$
$$x > -2$$

STUDY PROBLEM 4

$$-3(2m - 8) < 2(m + 8)$$
$$-6m + 24 < 2m + 16$$
$$-6m - 2m < +16 - 24$$
$$-8m < -8$$
$$m > +1$$

Dividing both sides by −8, the inequality sign is reversed.

STUDY PROBLEM 5

$$-3x+6(x-4)>10x+25$$
$$-3x+6x-24>10x+25$$
$$-3x+6x-10x>25+24$$
$$-7x>+49$$
$$x<-7$$

Dividing both sides by -7, the inequality sign is reversed.

STUDY PROBLEM 6

$$\frac{7m}{5}+\frac{13}{1}\leq\frac{3+m}{1}$$
$$\frac{(\cancel{5})7m}{\cancel{5}}+\frac{(5)13}{1}\leq\frac{(3+m)(5)}{1}$$

Steps:

1. LCD $= 5$

2. Multiply each numerator by 5

3. Divide each numerator by the denominator

4. Clear fractions and solve

$$7m+65\leq15+5m$$
$$7m-5m\leq15-65$$
$$2m\leq-50$$
$$m\leq-25$$

PRACTICE PROBLEMS: SOLVING INEQUALITIES

Solve each inequality.

1. $3t + 3 > 4t + 7$

2. $-7(3x + 4) > 35$

3. $-8(-2x - 3) + 5 \le 5(3x - 5)$

4. $\dfrac{(a+2)}{5} + \dfrac{(a-3)}{2} \le \dfrac{8}{1}$

ANSWERS TO PRACTICE PROBLEMS: SOLVING INEQUALITIES

1.
$$3t + 3 > 4t + 7$$
$$3t - 4t > 7 - 3$$
$$-t > 4$$
$$t < -4$$

4.
$$\dfrac{\overset{2}{(\cancel{10})}(a+2)}{\cancel{5}} + \dfrac{\overset{5}{(\cancel{10})}(a-3)}{\cancel{2}} \le \dfrac{(10)8}{1}$$

1. LCD = 10
2. Multiply each numerator by the LCD
3. Divide each numerator by the denominator of each fraction.
4. Solve the inequality.

$$2(a+2) + 5(a-3) \le 80$$
$$2a + 4 + 5a - 15 \le 80$$
$$7a - 11 \le 80$$
$$7a \le 91$$
$$a \le 13$$

2.
$$-7(3x + 4) > 35$$
$$-21x - 28 > 35$$
$$-21x > 35 + 28$$
$$-21x > 63$$
$$x < -3$$

3.
$$-8(-2x - 3) + 5 \le 5(3x - 5)$$
$$+16x + 24 + 5 \le 15x - 25$$
$$16x - 15x \le -25 - 24 - 5$$
$$1x \le -54$$

WORD PROBLEMS WITH INEQUALITIES

STUDY PROBLEM 8

A man in retirement found that he needed more than $2,500 per year return on his investments to live comfortably. If he invested $10,000 at 7%, how much must he invest at 9% to attain his goal?

Let x = amount invested at 9%
$.09x$ = return from 9% investment
$.07(10,000)$ = return from 7% investment

Equation: 7% investment $+$ 9% investment $> 2,500$

$$.09x + .07(10,000) > 2,500$$
$$9x + 70,000 > 250,000$$
$$9x > 180,000$$
$$x > 20,000$$

PRACTICE PROBLEMS:
WORD PROBLEMS WITH INEQUALITIES

1. The longer side of a triangle is 3 times the shorter side. The third side is 4 inches longer than the shorter. How long is the shorter side if the perimeter is more than 39 inches?

2. A teenager buys a sports jacket and a pair of jogging shorts. He decides to spend no more than $27 for both articles. If the sports jacket costs $6 more than twice the jogging shorts, how much did he spend for the jogging shorts?

3. A certain type of coolant contains a 65% efficiency rating against heat. Another type contains a 90% efficiency rating against heat. How many quarts of the 90% and 65% efficiency rating must be mixed together to get at least 8 quarts with an 80% efficiency rating?

ARCO ■ *Step-by-Step: How to Solve Algebra Word Problems*

ANSWERS TO PRACTICE PROBLEMS:
WORD PROBLEMS WITH INEQUALITIES

1. Let x = inches in shorter side

$3x$ = inches in longer side

$x + 4$ = inches in the third side

Equation: $x + 3x + x + 4 > 39''$

$5x + 4 > 39''$

$5x > 35$

$x > 7$ inches on the shorter side

2. Let x = cost of jogging shorts

$2x + 6$ = cost of sports jacket

Equation: $x + 2x + 6 \leq 27$

$3x \leq 27 - 6$

$3x \leq 21$

$x \leq \$7$

3.

Equation: $.90(x)+.65(8-x)\geq.80(8)$

Move decimal point 2 places to the right.

$$90x+520-65x\geq640$$

$$90x-65x\geq640-520$$

$$25x\geq120$$

$$25x\geq120 \text{ qt}$$

$$x\geq4\tfrac{4}{5}\text{ qt of the 90\% rating}$$

$$8-x\geq3\tfrac{1}{5}\text{ qt of the 65\% rating}$$

Chapter Fifteen

Solving Problems With Two Variables

In the previous chapters of this book, word problems were solved using one unknown, namely x, or any other single letter.

This chapter will show you how to solve word problems using two variables (x, y) instead of one. Using two variables gives you greater flexibility and allows you to solve more difficult problems.

Two variables means that we must establish two equations. The standard format for equations with two unknowns is as follows:

Both the x and y variables are placed in the left member. The constants are placed in the right member.

EXAMPLES:

a. $x = y + 10 \longrightarrow x - y = 10$

b. $2y = -2x - 15 \longrightarrow 2x + 2y = -15$

c.
$$2(y-x) = -3y + 16$$
$$2y - 2x = -3y + 16 \longrightarrow -2x + 5y = 16$$

d.
$$6(-x - y) = 3(3x + 8)$$
$$-6x - 6y = 9x + 24$$
$$-6x - 9x - 6y = 24 \longrightarrow -15x - 6y = 24$$

e.
$$\frac{3y}{7} - \frac{2x}{21} = 1$$
$$\text{L.C.D.} = 21$$
$$9y - 2x = 21 \longrightarrow -2x + 9y = 21$$

f.
$$-x - 6 = -y - 6$$
$$-x + y = -6 + 6 \longrightarrow -x + y = 0$$

g. $\dfrac{(2-x)}{2}+\dfrac{(y+3)}{4}=\dfrac{3(y-6)}{3}$

$LCD = 12$

$6(2-x)+3(y+3)=12(y-6)$

$12-6x+3y+9=12y-72$

$-6x+3y-12y=-72-21 \longrightarrow -6x-9y=-93$

PRACTICE PROBLEMS: TWO VARIABLES
(SETTING UP THE EQUATION)

1. $x = -y + 16$

2. $3(-4x + 2) = -y$

3. $\dfrac{(x+y)}{6}-\dfrac{x}{12}=\dfrac{3(-x-y)}{6}$

ANSWERS TO PRACTICE PROBLEMS: TWO VARIABLES
(SETTING UP THE EQUATION)

1. $x + y = 16$

2. $-12x + y = -6$

3. $7x + 8y = 0$

STUDY PROBLEM 1

The sum of two numbers is 15 and their difference is 7. What are the two numbers?

Let $x =$ larger number

$y =$ smaller number

Equations: (1) $x + y = 15$

(2) $x - y = 7$

By adding both equations (1) and (2) we have

$$2x = 22$$
$$x = 11 \text{ (larger number)}$$

Substituting 11 for x in equation (1) we have

$$11 + y = 15$$
$$+y = 15 - 11$$
$$y = 4 \text{ (smaller number)}$$

Solution Set (11, 4)

RULE

To eliminate the same variable of both equations, the signs of each must be the inverse of the other and their coefficients must be the same.

STUDY PROBLEM 2

Equation (1) $x + y = 8$
Equation (2) $x - 2y = 2$

Notice that the x variables in both equations have the same coefficients (1) and the same sign (+). If we change the sign of either equation (1) or equation (2), the x variable will be eliminated. Here we can use the subtraction method.

Subtraction method: (1) $- x - y = -8$
(2) $\underline{x - 2y = 2}$
$$-3y = -6$$
$$y = 2$$

Notice that if we reverse the sign of x to $-x$, we must also reverse the signs of the other terms in the equation.

Substituting 2 for y in equation (2), we have

$$x - 4 = 2$$
$$x = 6$$

Solution Set (6, 2)

STUDY PROBLEM 3

$$\text{Equation (1)} \quad x + 2y = 8$$
$$\text{Equation (2)} -2x + 2y = -10$$

Using subtraction method

$$x = 6$$

Substituting 6 for x in equation (1)

$$6 + 2y = 8$$
$$2y = 2$$
$$y = 1$$

Solution Set (6, 1)

STUDY PROBLEM 4

Caren's petty cash contains 100 coins made up of nickels and quarters. How many nickels and quarters does she have in her petty cash if the worth of all coins is $17?

Let x = number of quarters
Let y = number of nickels

$$\text{Equation (1)} \quad x + y = 100 \text{ coins}$$
$$\text{Equation (2)} \, .25x + .05y = 17.00$$

Move decimal point 2 places to the right.

$$(1) \qquad x + y = 100$$
$$(2) \qquad 25x + 5y = 1,700$$

Multiplying equation (1) by -5

$$(1) \quad -5x - 5y = -500$$
$$(2) \quad \underline{25x + 5y = 1,700}$$

Adding (1) and (2)
$$20x = 1,200$$
$$x = 60 \text{ quarters}$$
$$y = 40 \text{ nickels}$$

As you can see above, there are *three methods* for eliminating the same variable of both equations:

1. addition

2. subtraction

3. multiplication

PRACTICE PROBLEMS: TWO VARIABLES
(ADDITION AND SUBTRACTION METHODS)

Using the addition method, solve for x and y in Problems 1 through 3.

1. (1) $x + y = 12$

    (2) $x - y = 4$

2. (1) $6x - 5y = 9$

    (2) $2x + 5y = 23$

3. (1) $.5x + .7y = 6.4$

    (2) $.2x - .7y = .6$

Place in standard format and then use the addition method to solve Problem 4.

4. (1) $3(x + 1) = 2(y + 1)$

    (2) $2y + 3 = 2(x + 4)$

Using the subtraction method, solve for x and y in Problems 5 through 7.

5. (1) $x + 2y = 16$

    (2) $2x + 2y = 24$

6. (1) $-x + 3y = 6$

    (2) $-x + y = 10$

7. (1) $.08x - .05y = 6$

    (2) $.08x - .02y = 3$

Rearrange to standard format and then use the subtraction method to solve Problem 8.

8. (1) $3(x+2y)=-2x+8$

(2) $2(y+3)=x-4y-10$

ANSWERS TO PRACTICE PROBLEMS: TWO VARIABLES (ADDITION AND SUBTRACTION METHODS)

Solution sets for the addition method.

1. (8, 4)

2. (4, 3)

3. (10, 2)

4. $(4, 6\frac{1}{2})$

Solution sets for the subtraction method.

5. (8, 4)

6. (−12, −2)

7. $(12\frac{1}{2}, -100)$

8. (4, −2)

STUDY PROBLEM 5

Use the multiplication method to eliminate the same variable of each equation. Then solve for x and y.

$$(1)\quad x+2y=-1$$
$$(2)\ 5x-4y=16$$

Multiplying equation (1) by 2 we get

$$(1)\ 2x+4y=-2$$
$$(2)\ \underline{5x-4y=16}$$
$$7x\quad\ =14$$
$$x=2$$

Substituting 2 for x in equation (1)

$$(1)\ 2+2y=-1$$
$$2y=-1-2$$
$$2y=-3$$
$$y=-3/2=-1\tfrac{1}{2}$$

Solution Set is $(2,\ -1\tfrac{1}{2})$

STUDY PROBLEM 6

Use the multiplication method to eliminate the same variable of each equation. Then solve for x and y.

$$(1)\quad 4x+2y=0$$
$$(2)\quad x-5y=-11$$

Multiplying equation (2) by -4 we get

$$(1)\quad\ 4x+2y=0$$
$$(2)\ -4x+20y=+44$$
$$\overline{22=+44}$$
$$y=+2$$

Substituting 2 for y in equation (1)

$$(1)\quad\ 4x+4=0$$
$$4x=-4$$
$$x=-1$$

Solution Set is $(-1, 2)$

STUDY PROBLEM 7

Use the multiplication method to eliminate the same variable of each equation. Then solve for x and y.

$$(1) \quad 3x - 8y = -18$$
$$(2) \quad 2x + 5y = 19$$

Multiplying equation (1) by –2 and equation (2) by +3 we get

$$(1) \quad -6x + 16y = 36$$
$$\underline{(2) \quad 6x + 15y = 57}$$
$$31y = 93$$
$$y = 3$$

Substituting 3 for y in equation (1)

$$(1) \quad 3x - 24 = -18$$
$$3x = 24 - 18$$
$$3x = 6$$
$$x = 2$$

Solution Set is (2, 3)

PRACTICE PROBLEMS: TWO VARIABLES
(MULTIPLICATION METHOD)

Using the multiplication method, solve for x and y in the following problems.

1. (1) $2x + 3y = 23$

  (2) $3x + y = 17$

2. (1) $5x - 3y = 23$

  (2) $7x - 4y = 33$

Rearrange to standard format and find the solution set.

3. (1) $-1-3y = 2(y-2x)$

(2) $2(x-y) = 2$

ANSWERS TO PRACTICE PROBLEMS: TWO VARIABLES (MULTIPLICATION METHOD)

1. (1) $2x + 3y = 23$

(2) $3x + y = 17$

Multiplying (2) by -3

$$
\begin{array}{rl}
(1) & -9x - 3y = -51 \\
 & \underline{2x + 3y = 23} \\
 & -7x = -28 \\
 & x = +4
\end{array}
$$

Substituting 4 for x in (1)

$$
\begin{array}{l}
8 + 3y = 23 \\
3y = 23 - 8 \\
3y = 15 \\
y = 5
\end{array}
$$

Solution Set (4, 5)

2. (1) $5x - 3y = 23$

(2) $7x - 4y = 33$

Multiplying (1) by −4 and (2) by 3

$$(1) \quad -20 + 12y = -92$$
$$(2) \quad \underline{21x - 12y = 99}$$
$$1x = 7$$

Substituting 7 for x in (1)

$$(1) \quad 35 - 3y = 23$$
$$-3y = -12$$
$$y = +4$$

Solution Set (7, 4)

3. Rearranged

$$(1) \quad 4x - 5y = 1$$
$$(2) \quad 2x - 2y = 2$$

Multiplying (2) by −2

$$(2) \quad -4x + 4y = -4$$
$$\underline{4x - 5y = 1}$$
$$-1y = -3$$
$$y = +3$$

Substituting 3 for y in (1)

$$4x - 15 = 1$$
$$4x = 16$$
$$x = 4$$

Solution Set (4, 3)

FRACTIONAL EQUATIONS CONTAINING TWO VARIABLES

When one or both of the equations to be solved contains a fraction, you must find the lowest common denominator and clear the fractions before eliminating one variable and solving for x and y.

STUDY PROBLEM 8

$$\text{(1)} \qquad \frac{x}{2} + \frac{y}{3} = 2$$

$$\text{(2)} \qquad \frac{x}{3} + \frac{y}{9} = 1$$

Equation (1) LCD = 6 (multiply each fraction by 6)

$$6\left(\frac{x}{2}\right) + 6\left(\frac{y}{3}\right) = 6\left(\frac{2}{1}\right)$$
$$3x + 2y = 12$$

Equation (2) LCD = 9 (multiply each fraction by 9)

$$9\left(\frac{x}{3}\right) + 9\left(\frac{y}{9}\right) = 9\left(\frac{1}{1}\right)$$
$$3x + y = 9$$

Subtract Equation (1) from Equation (2)

$$3x + y = 9$$
$$\underline{2x + 2y = 12}$$
$$-y = -3$$
$$y = +3$$

Substituting (+3) for y in equation (2)

$$(2) \qquad 3x + 3 = 9$$
$$3x = 6$$
$$x = 2$$

Solution Set (2, 3)

PRACTICE PROBLEMS: TWO VARIABLES (ANY METHOD)

In the systems of equations on pages 113–117, use any of the three convenient methods to find the solution set.

1. (1) $3x - 5y = 4$
   (2) $2x + 5y = 16$

2. (1) $-7x + y = 2$
   (2) $7x + 2y = 4$

3. (1) $-3x + 2y = 6$
   (2) $-4x - 2y = -20$

4. (1) $2x + 3y = 4$
   (2) $x + 3y = 14$

5. (1) $11x - y = 17$
   (2) $7x - 4y = -6$

6. (1) $-q = -8p + 22$
   (2) $7p = q + 18$

7. (1) $2(x+y)+10 = 5(x+1)$

(2) $x+y-2 = 2(y-1)$

8. (1) $\dfrac{x}{2} - \dfrac{3y}{4} = -2$

(2) $\dfrac{y}{2} + \dfrac{x}{4} = \dfrac{4}{1}$

Check your answers below.

ANSWERS TO PRACTICE PROBLEMS: TWO VARIABLES

1. $(4,\ 1\tfrac{3}{5})$

2. $(0, 2)$

3. $(2, 6)$

4. $(-10, 8)$

5. $(2, 5)$

6. $(4, 10)$

7. $(5, 5)$

8. $(4\tfrac{4}{7}, 5\tfrac{5}{7})$

WORD PROBLEMS CONTAINING TWO VARIABLES

STUDY PROBLEM 9

Bob wants to create a special blend of tea by mixing tea costing $1.65 per pound with tea costing $1.85 per pound. To make up 7 pounds of the special blend at a cost of $1.75 per pound, how many pounds of each brand of tea should he use?

Model:

Since we have *two* unknowns, x and y, we must generate *two* equations. These equations we will call (1) *sum* and (2) *product*. The sum equation deals with pounds.

$$(1)\ x\ \text{lb} + y\ \text{lb} = 7\ \text{lb}$$

The product deals with cost.

$$(2)\ \$1.65x + \$1.85y = \$12.25$$

$$\begin{aligned}
&\text{Equation (1)} && x + y = 7 \\
&\text{Equation (2)}\ && 100(1.65x + 1.85y) = 100(12.25)
\end{aligned}$$

Multiplying by 100 moves the decimal point 2 places so that we now have whole numbers.

$$165x + 185y = 1{,}225$$

$$\begin{aligned}
&\text{Equation (1)} && x + y = 7 \\
&\text{Equation (2)}\ && 165x + 185y = 1{,}225
\end{aligned}$$

Our next step is to eliminate the x's by multiplying the equation (1) by –165.

$$-165(x+y)=-165\cdot7$$

Equation (1) $-165x-165y=-1,155$

Equation(2) $\underline{165x+185y=1,225}$

By addition $20y=70$

$$y=3\tfrac{1}{2}\text{ lb}$$

$$x+3\tfrac{1}{2}=7$$

Substituting in Equation (1) $x=3\tfrac{1}{2}\text{ lb}$

STUDY PROBLEM 10

Susan's piggy bank contains 96 coins made up of nickels and quarters. How many nickels and quarters does she have in the piggy bank if the worth of all coins is $12.00?

Let x = number of nickels
Let y = number of quarters

Sum equation (1) $x + y = 96$ (number of coins)

Let $5x$ = value of the nickels in pennies
Let $25y$ = value of the quarters in pennies

Product equation (2) $5x + 25y = 1,200$ (total value of coins in pennies)

Multiplying equation (1) by –5

(1) $-5x-5y=-480$

(2) $\underline{5x+25y=1,200}$

By addition $20y=720$

$$y=36\text{ quarters}$$

Substituting in Equation (1) $x=60\text{ nickels}$

STUDY PROBLEM 11

A man has invested $15,000, part in 7% stocks and the remainder in 9% AA Bonds. If his yearly yield from both investments is $1230, how much did he invest at each rate?

Model:

(1) sum	$\dfrac{\text{Stocks}}{x}$	$+$	$\dfrac{\text{AA Bonds}}{y}$	$=$	$\dfrac{\text{15,000 Invested}}{\text{Annual Yield}}$
(2) product	7%		9%		$1,230.00

$$\text{Sum equation (1)} \qquad x + y = \$15,000$$
$$\text{Product equation (2)} \quad 7\%x + 9\%y = \$1,230.00$$

Multiplying (2) by 100

$$(2) \quad .07x + .09y = \$1,230.00$$
$$(2) \qquad 7x + 9y = 123,000$$
$$(1) \qquad x + y = 15,000$$

Multiplying (1) by –7

$$(1) \quad -7x - 7y = -105,000$$
$$(2) \quad \underline{7x + 9y = 123,000}$$

Addition
$$2y = 18,000$$
$$y = \$9,000 \text{ invested in 9\% AA Bonds}$$

Substituting in (1)
$$x = \$6,000 \text{ invested in 7\% stocks}$$

STUDY PROBLEM 12

A pharmacist has a 12% solution of boric acid and a 20% solution of boric acid. How many grams of the 12% and 20% solution must he use to make 100 grams of a 15% solution?

Model:

Sum equation (1) $x + y = 100$

Product equation (2) $.12x + .20y = .15(100)$

$12x + 20y = 1,500$

Multiplying (1) by −12 we get

(1) $-12x - 12y = -1,200$

(2) $\underline{12x + 20y = 1,500}$

Add (1) and (2) $8y = 300$

$y = 37\frac{1}{2}$ grams of 20% solution

Substituting in (1) $x = 62\frac{1}{2}$ grams of 12% solution

STUDY PROBLEM 15

An airplane makes a test run with a tail wind and records a speed of 682 miles per hour. On the return trip against the same wind, the plane can attain only 624 miles per hour. What is the speed of the plane in still air and what is the wind's velocity?

Let x = speed of plane in still air
Let y = wind's velocity

With tail wind ⟶

Equation (1) $x + y = 682$

⟵ Return trip against wind

Equation (2) $x - y = 624$

By addition of (1) and (2)

$$2x = 1,306$$
$$x = 653 \text{ (speed of plane)}$$

By substitution $y = 29$ (wind's velocity)

STUDY PROBLEM 16

Two planes start at the same time and fly in opposite directions. The faster plane travels twice as fast as the slower plane. After 3 hours, they are 2,520 miles apart. How fast does each plane travel?

Let x = speed of the fast plane
Let y = speed of the slow plane

Equation (1) $x = 2y$
Equation (2) $3x + 3y = 2,520$
(1) $x - 2y = 0$
(2) $3x + 3y = 2,520$

Multiply by –3

(1) $-3x + 6y = 0$
(2) $3x + 3y = 2,520$

By addition $9y = 2,520$

$y = 280$ mph (slower plane)

By substitution $x = 560$ mph (faster plane)

STUDY PROBLEM 17

Four hundred tickets were sold for a school play, bringing in a total of $1025. If adults were charged $3.50 and students $2, how many adults and how many students attended the performance?

Model:

(1) sum	$\dfrac{x}{3.50x}$	$+$	$\dfrac{y}{2.00y}$	$=$	400 attendance
(2) product					1,025 collected

Let x = adult tickets sold
Let y = student tickets sold

$$(1) \qquad\qquad x + y = 400$$
$$(2) \quad 3.50x + 2.00y = 1,025.00$$
$$(3) \quad 350x + 200y = 102,500$$

Multiplying (1) by −200

$$(1)\ \ -200x - 200y = -80,000$$

By addition $\qquad (2) \qquad\qquad 150x = 22,500$

$$x = 150 \text{ adult tickets}$$

By substitution $\qquad\qquad\qquad y = 250 \text{ student tickets}$

PRACTICE PROBLEMS:
WORD PROBLEMS USING TWO VARIABLES

Check your answers with the solutions on pages 134–139.

1. The sum of two numbers is 48 and their difference is 12. What are the two numbers?

2. The sum of 2 numbers is 18. Three times the larger less the smaller equals 22. What are the numbers?

3. The perimeter of a rectangle is 120 cm. Three times its width equals the length. What are the dimensions?

4. There are 24 coins made up of quarters and dimes which amount to $3.90. How many of each kind of coin are there?

5. Mr. Owens, a financial consultant, recommended that a client invest $15,000, part at 7% and the remainder at 9%, to insure a $1,300.00 yearly income. How much was invested at each rate?

6. One dozen oranges and 6 grapefruit cost $2.46. Nine oranges and 9 grapefruit cost $2.97. Find the cost of 1 orange and 1 grapefruit.

7. Euclid High School received a shipment of algebra and geometry books costing $4200. The cost of 1 algebra book is $12 and the cost of 1 geometry book is $15. If there were 125 more algebra books than geometry books, how many of each kind of math book were in the shipment?

8. A merchant has two kinds of Christmas candy, one selling at $2.20 a pound and the other at $.95 a pound. How many pounds of each should he mix to sell 100 pounds at $2.10?

9. A labor union has $100,000 to invest for its membership. It was decided that the investment would be in an 8% preferred stock and a 12% AA Bond. If the yearly income from both investments is $10,400, how much was invested in the stocks and in the AA Bonds?

10. If 1 pound of coffee and 8 pounds of sugar cost $9.11, and 5 pounds of coffee and 5 pounds of sugar cost $14.40, find the cost of 1 pound of coffee.

11. Ralph Peterson, a graduating senior at U.S.C., received an annuity amounting to $14,800 from his grandparents. He quickly invested part of it in Treasury Bonds giving a 7% return and the remainder in preferred stocks yielding a 6½% yearly return. If he expects $1002 from both investments, what amount did he invest in the Treasury Bonds and what amount in preferred stocks?

12. If Darlene can purchase 6 cans of peaches and 12 cans of tomatoes for $11.10 or 12 cans of peaches and 6 cans of tomatoes for $10.86, what is the price of 1 can of peaches and 1 can of tomatoes?

13. Mr. Blossom, a farmer, paid 10 men and 8 boys $970 for 1 day's work. Later he paid 12 men and 6 boys $1020 for a day's work. The men were paid at a uniform wage and the boys were paid at another uniform wage. What daily wage was paid to the men and boys?

14. A trust fund containing $30,000 had been established for a young adult. Part of the money was invested in a Savings and Loan Association which paid a 5½% yearly yield and the remainder was invested in Second Trust Deeds yielding a 7% return. If he received $1830 from the two investments annually, how much did he invest in each?

15. Two cars started from the same point but traveled in opposite directions. The faster car traveled 15 kilometers faster than the slower car. After five hours the 2 cars were 575 kilometers apart. How many kilometers per hour did each car travel?

16. A softball game was planned by the Rotary Club to raise funds to build an athletic facility for handicapped persons. Adults other than senior citizens were charged 5 dollars. All others were charged 2 dollars each. If the total attendance was 3,702 and if the sum of 16,404 dollars was collected, how many adult tickets and all other tickets were sold?

17. An F-18 fighter pilot makes a test run from Andrews Air Force Base in California to La Guardia Airport in New York City averaging 770 miles per hour with a tail wind. On the return trip against the same wind, the plane averaged 680 miles per hour. What is the speed of the plane in still air and what is the velocity of the wind?

18. If 2 high school football players with a combined weight of 396 pounds balance a lever when they sit respectively 6 feet and 5 feet from the fulcrum, what is each player's weight?

19. At a famous university the alumni of the School of Liberal Arts and the School of Engineering decided they would together raise $545,000 for their Alma Mater to commemorate the 25th anniversary of their graduation. There were a total of 4,200 pledges. The Liberal Arts school averaged $150 per pledge and the School of Engineering averaged $100 per pledge. How many alumni from each school made pledges?

20. One-fifth of the sum of Lorna's age in 5 years is equal to $\frac{1}{4}$ of Larry's age less 1. Two-fifths of the sum of Larry's age in 4 years is equal to $\frac{1}{5}$ of twice Lorna's age plus 2. What are their ages now?

Hint: Let x = Lorna's age

   Let y = Larry's age

   Find LCD of each equation and then clear fractions

ANSWERS TO PRACTICE PROBLEMS:
WORD PROBLEMS USING TWO VARIABLES

1. $x + y = 48$

$\underline{x - y = 12}$

$2x = 60$

$x = 30$

$y = 18$

2. $x + y = 18$

$\underline{3x - y = 22}$

$4x = 40$

$x = 10$

$y = 8$

3. (Multiply by 2)

$2L + 2W = 120$

$3W = L$

$2L + 2W = 120$

$-L + 3W = 0$

$2L + 2W = 120$

$\underline{-2L + 6W = 0}$

$8W = 120$

$W = 15$

$L = 45$

4. (Multiply by −10)

$q + d = 24$

$25q + 10d = 390$

$-10q - 10d = -240$

$\underline{25q + 10d = 390}$

$15q = 150$

$q = 10$

$d = 14$

5. (Multiply by –7)

$$x + y = 15,000$$
$$7x + 9y = 130,000$$
$$-7x - 7y = -105,000$$
$$7x + 9y = 130,000$$
$$2y = 25,000$$
$$y = 12,500 \ @ \ 9\%$$
$$x = 2,500 \ @ \ 7\%$$

6. (Multiply by -3) $12x + 6y = \$2.46$
   (Multiply by -2) $9x + 9y = \$2.97$

$$-36x - 18y = -\$7.38$$
$$18x + 18y = \$5.94$$
$$-18x = -\$1.44$$
$$x = 8 \text{ cents cost of 1 orange}$$
$$y = 25 \text{ cents cost of 1 grapefruit}$$

$$12x + 15y = 4,200$$
7. (Multiply by -12) $x - y = 125$

$$12x + 15y = 4,200$$
$$-12x + 12y = -1,500$$
$$27y = 2,700$$
$$y = 100 \text{ geometry books}$$
$$x = 225 \text{ algebra books}$$

8. (Multiply by –95)

$$x + y = 100$$
$$95x + 220y = 21,000$$
$$-95x - 95y = -9,500$$
$$95x + 220y = 21,000$$
$$125y = 11,500$$
$$y = 92 \text{ lb} \ @ \ \$2.20$$
$$x = 8 \text{ lb} \ @ \ \$.95$$

9. (Multiply by −8)

$$x + y = 100,000$$
$$8x + 12y = 1,040,000$$
$$-8x - 8y = -800,000$$
$$\underline{8x + 12y = 1,040,000}$$
$$4y = 240,000$$
$$y = \$60,000 \text{ invested in 12\% AA Bonds}$$
$$x = \$40,000 \text{ invested in 8\% preferred stock}$$

10. (Multiply by −5)

$$1x + 8y = 911$$
$$5x + 5y = 1,440$$
$$-5x - 40y = -4,555$$
$$\underline{5x + 5y = 1,440}$$
$$-35y = -3,115$$
$$y = \$.89 \text{ sugar per lb}$$
$$x = \$1.99 \text{ coffee per lb}$$

11. (Multiply by −7)

$$x + y = 14,800$$
$$7x + 6\tfrac{1}{2}y = 100,200$$
$$-7x - 7y = -103,600$$
$$\underline{7x + 6\tfrac{1}{2}y = 100,200}$$
$$-\tfrac{1}{2}y = -3,400$$
$$y = \$6,800 \text{ invested in preferred stock}$$
$$x = \$8,000 \text{ invested in Treasury bonds}$$

12. (Multiply by –2)

$$6x + 12y = 11.10$$
$$12x + 6y = 10.86$$
$$-12x - 24y = -22.20$$
$$12x + 6y = 10.86$$
$$\overline{}$$
$$-18y = -11.34$$
$$y = \$.63 \text{ cost of 1 can of tomatoes}$$
$$6x + 7.56 = 11.10$$
$$6x = 3.54$$
$$x = \$.59 \text{ cost of 1 can of peaches}$$

13. (Multiply by 3)
(Multiply by -4)

$$10x + 8y = 970$$
$$12x + 6y = 1,020$$
$$30x + 24y = 2,910$$
$$-48x - 24y = -4,080$$
$$\overline{}$$
$$-18x = -1,170$$
$$x = \$65 \text{ men's daily wage}$$
$$650 + 8y = 970$$
$$8y = 970 - 650$$
$$8y = 320$$
$$y = \$40 \text{ boy's daily wage}$$

$$x + y = 30,000$$

14. (Multiply by –7)

$$5\tfrac{1}{2}x + 7y = 183,000$$
$$-7x - 7y = -210,000$$
$$5\tfrac{1}{2}x + 7y = 183,000$$
$$\overline{\phantom{5\tfrac{1}{2}x + 7y = 183,000}}$$
$$-1\tfrac{1}{2}x = -27,000$$
$$-\tfrac{3}{2}x = -27,000$$
$$x = \$18,000 @ 5\tfrac{1}{2}\%$$
$$y = \$12,000 @ 7\%$$

15. Let x = rate of faster car

Let y = rate of slower car

$$x - y = 15 \text{ km}$$
(Multiply by –5) $\qquad 5x + 5y = 575 \text{ km}$
$$-5x + 5y = -75$$
$$\underline{5x + 5y = 575 \text{ km}}$$
$$10y = 500$$
$$y = 50 \text{ km rate of slower car per hour}$$
$$x = 65 \text{ km rate of faster car per hour}$$

16. Let x = number of adults attending

Let y = number of "others" attending

$$x + y = 3,702$$
(Multiply by –5) $\qquad 5x + 2y = 16,404$
$$-5x - 5y = -18,510$$
$$\underline{5x + 2y = -16,404}$$
$$-3y = -2,106$$
$$y = 702 \text{ "others"}$$
$$x = 3,000 \text{ adults}$$

17. Let x = speed in still air

Let y = wind velocity

$$x + y = 770$$
$$\underline{x - y = 680}$$
$$2x = 1,450$$
$$x = 725 \text{ mph in still air}$$
$$y = 45 \text{ mph wind velocity}$$

18. Let x = weight of first football player

Let y = weight of second football player

$$x + y = 396$$
$$6x = 5y$$

(Multiply by 5) $x + y = 396$

$$6x - 5y = 0$$
$$5x + 5y = 1,980$$
$$\underline{6x - 5y = 0}$$
$$11x = 1,980$$
$$x = 180 \text{ lbs of one player}$$
$$y = 216 \text{ lbs of other player}$$

19. Let x = number of Liberal Arts pledges

Let y = number of Engineering School pledges

$$x + y = 4,200$$

(Multiply by –100) $150x + 100y = 545,000$

$$-100x - 100y = -420,000$$
$$\underline{150x + 100y = 545,000}$$
$$50x = 125,000$$
$$x = 2,500 \text{ Liberal Arts pledges}$$
$$y = 1,700 \text{ Engineering School pledges}$$

20. (Multiply by 2)

$$\frac{x+5}{5} = \frac{y}{4} - \frac{1}{1}$$
$$\frac{2}{5}(y+4) = \frac{2x}{5} + 2$$
$$-2x + 2y = 2$$
$$4x - 5y = -40$$
$$\underline{-4x + 4y = +4}$$
$$-y = -36$$
$$y = 36 \text{ Larry's age}$$
$$x = 35 \text{ Lorna's age}$$

Chapter Sixteen

Word Problems Review Test

Use one or two variables to solve each of the following problems. Check your answers with the solutions on pages 147–170.

1. The length of a rectangle is 10 less than 3 times the width. If the perimeter is 140 cm, what are the dimensions of the rectangle?

2. Angle A is 3 times angle B and angle B is twice angle C. How many degrees are there in each angle of triangle ABC? (There are 180 degrees in every triangle.)

3. The length of a rectangle is 10 less than the width. If the perimeter is 140 centimeters, what are the dimensions of the rectangle?

4. John has 98 coins in his bank made up of dimes and quarters. The total value of the coins is $18.50. How many of each coin does he have?

5. Frank is 4 years older than John. Five times Frank's age 2 years from now will equal 7 times John's age 2 years ago. What are their present ages?

6. In 4 years, Nancy will be as old as Jane is now. Seven years ago the sum of their ages was 18. Find their ages.

7. A woman has 2 investments which yield an annual income of $996. One investment yields 7% and the other 8%. How much is invested in each if the total amount invested is $12,800?

8. Mr. Rosenberg invested part of his $40,000 investment in high yield bonds paying $9\frac{1}{2}\%$ and preferred stock paying 8%. If his annual income from both investments is $3620, how much did he invest in the bonds and in the stocks?

9. How many ounces of an alloy containing 35% aluminum must be melted with an alloy containing 65% aluminum to obtain 30 ounces of an alloy containing 55% aluminum?

10. Mr. Washburn, a plumber, can complete a job in 6 hours and his helper can do the same job in 9 hours. If they both work together to complete the job, how many hours will it take them?

11. A computer expert can do a job on the computer in 6 hours. His secretary can do the same job in 7 hours. After the expert works on the job for 2 hours, his secretary takes over and completes the job. How many hours did it take her to complete the job?

12. How many quarts of pure water must be added to 15 quarts of a solution that is 15% acid in order to dilute it to a solution that is only 12% acid?

13. Two pipes can fill a tank in 10 hours and 12 hours respectively. A third pipe can drain the tank in 15 hours. How long will it take the tank to be filled if all three pipes are open?

14. An experienced typist can type a legal brief in 6 hours. Her assistant can type the same brief in 9 hours. If they both work together for 3 hours and then the experienced typist works alone, how many hours will it take her to finish the job?

15. Two routes to get from Ventura, California to San Diego, California differ by 20 miles. The time required is the same if a car travels 50 miles per hour over the shorter route and 55 miles per hour over the longer route. How long is each route in miles?

16. Two motorcyclists start toward each other from towns 318 miles apart. If one motorcyclist travels 52 miles per hour and the other 54 miles per hour, in how many hours will they meet?

17. If a weight of 130 pounds is 14 inches from the fulcrum, how far must a weight of 140 pounds be placed from the fulcrum to maintain balance?

18. Two tuna boats leave the same port traveling in opposite directions along the west coast of the United States. One boat travels 5 miles per hour faster than the other. At the end of one day's travel, they are 1,080 miles apart. What is the speed of each boat per hour?

19. The charge for admission to the natural history museum is $1.50 for adults and 1 dollar for children. If 5,000 people visited the museum in one day and the total receipts were $6500, how many of each kind of ticket was sold?

20. A mason contractor needed to build a brick warehouse in 2 days. The first day he hired 8 bricklayers and 4 apprentices at a cost of $1280 per day. The second day he hired 10 bricklayers and 6 apprentices at a cost of $1680 per day. What did it cost him per day to pay each bricklayer and each apprentice?

21. Mr. Randolph, a businessman, flew his jet plane from Newport Beach, California to San Francisco, a distance of 480 miles, in $1\frac{1}{2}$ hours against the wind. On his return trip, he covered the same distance in 1 hour. If the speed of the wind was the same going and returning, what was the speed of the tailwind?

22. The perimeter of a triangle is 173 centimeters. The first side is 10 inches greater than the second side. The third side is 5 inches less than twice the first side. Find the length of each side of the triangle.

23. Suzanne is one-sixth as old as her mother. In 10 years, she will be three-eighths as old as her mother will be then. What are their present ages?

24. Two giant wrestlers weighed a total of 450 pounds. To make a certain weight division, they both decided to go on a diet. One wrestler gained 15% of his weight and the other lost 10% of his weight. If the sum of their weights is now 461.25 pounds, how many pounds did each weigh before starting his diet?

25. Some oil companies are now using power recovery turbines that capture vast amounts of energy that previously went up the stack as waste gas. This device would save approximately 42 million gallons of gas per year per company. If the number of the present oil companies were increased by one-fifth less 2, the number of oil companies would total 22.
   a) How many oil companies are there presently?
   b) How many gallons of gas will be saved if the new companies use this energy saving device?
   c) How many gallons of gasoline would be saved with just the present number of oil companies using this device?

26. A grocer wishes to combine one grade of coffee worth $2.10 per pound with another grade worth $1.90 per pound to obtain 300 pounds worth $2 per pound. How many pounds of each grade did he mix?

27. A businessman invested part of $40,000 in stocks paying 6% and the remainder in bonds paying 7%. If his yearly income from both investments is $2720, how much did he invest in stocks and in bonds?

28. A total of $40,000 is invested, part at $4\frac{1}{2}\%$ and the other part at 6%. The income from both investments is $2130. How much was invested at each rate?

29. Two cyclists leave the same place going in the same direction. One cyclist travels 30 miles per hour and the second travels 35 miles per hour. If the second cyclist starts 2 hours later than the first, in how many hours will the second overtake the first?

30. Two motorbikes leave Atlanta, Georgia at the same time, one traveling north at the speed of 50 miles per hour and the other south at the rate of 55 miles per hour. After how many hours will they be 420 miles apart?

31. Two speedboats leave the port of San Francisco traveling in opposite directions at the same time. One boat travels 8 knots per hour faster than the other. After one day's travel they are 1,920 nautical miles apart. What are the speeds of the 2 boats?

32. In a walkathon, two men 27.5 miles apart start walking toward each other. One travels 10 miles in 4 hours and the other travels at the rate of 6 miles in 2 hours. How many miles will each have traveled when they meet?

33. A rocket travels twice the speed of a supersonic plane. If at the end of 1 hour they are 4,680 miles apart, find the speed of each if both flew in opposite directions from the same air field.

34. A mason, working alone, can tile a bathroom in 3 days. He has a helper who can tile the same job alone in 6 days. If the mason and his helper work together, how many days will it take them to do the job?

35. John can mow a front and back lawn in 3 hours. When his brother Bill helps him, together they can complete the mowing in 2 hours. How long will it take Bill to do the mowing alone?

36. One large pipe can fill a swimming pool in 10 hours and another pipe can drain the pool in 15 hours. If both pipes are opened, how many hours will it take before the pool is filled?

37. Ralph can paint a house in 4 days. His brother Walter can paint the same house in 6 days. After working 1 day, Ralph became ill and Walter finished the job. How long did it take Walter to finish the job?

38. A storage pipe can fill a reservoir in 15 days. Another smaller pipe can fill it in 20 days. After the larger pipe is on for 2 days, it breaks down. The smaller pipe is opened and completes the job of filling the reservoir. How long did it take for the smaller pipe to complete the job alone?

39. A Houston-based firm experimenting with gene splicing needs to increase its office space one and one-half times its present space. If they need 500,000 square feet of space, how many square feet do they have at the present and by how many square feet must they increase their space?
Let x = number of square feet at the present
  ? = number of square feet to be added

40. In 1982, the American car manufacturers suffered a decline in car sales that amounted to $\frac{1}{6}$ of the cars sold in 1979 less 3,000. If the number of cars sold in 1979 less the decline in car sales in 1982 amounted to 300,000 units, what was the decline in car sales in 1982? What were the car sales in 1979?

41. A private environmental protection agency made a study which ascertained that the air quality would be substantially improved in Los Angeles if there were 10,000 less than $\frac{1}{5}$ of the present number of cars on the roads. If the present number of cars on the roads less those which should not be on the roads is 480,000 units, how many cars are on the roads and by what amount should they be reduced?
Hint: Let x = number of cars present on the roads
    ? = number of cars that should be reduced

42. A tennis court has a perimeter of 212 feet. If the length is 24 feet longer than the width, what are the dimensions of the court? (Draw a rectangle.)

43. The radius of the larger circle $\overline{OB}$ is 14 feet and the radius of the smaller circle $\overline{OC}$ is 7 feet. What is the area of the shaded part? (Use $\pi = 22/7$. Area of the circle $= \pi r^2$.)

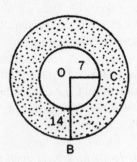

44. A navy boat goes 50 miles downstream in $2\frac{1}{2}$ hours. Going upstream it traveled the same distance in 4 hours. What is the rate of the boat in still water, and what is the rate of the current? (Draw model.)

45. A Federal Railways bus leaves Cleveland for Springfield and averages 56 miles per hour. Thirty minutes later a second bus leaves from the same station for Springfield averaging 60 miles per hour. How long will it take the second bus to overtake the Federal Railways bus?

46. An attendant at a service station checks a truck radiator and finds it to contain 40% antifreeze. If the radiator holds 20 quarts and is full, how many quarts must be drained off and replaced in order to get a 50% antifreeze reading?

47. The sum of the ages of two men is 62 years. One-half of the younger's age 6 years from now is equal to one-third of the older's age 2 years from now. How old is each man now?

Let x = age of older man
Let y = age of younger man

48. How many degrees in each angle of the triangle?

49. A worn-out pump can empty a tank in 8 hours. A new pump can empty the tank in 4 hours. After the old pump is used for 3 hours, the new pump replaces it and empties the tank. How long did it take for the new pump to empty the tank? (Draw model.)

50. A bricklayer can build a wall in 8 days. His helper can build the same wall in 10 days. After the bricklayer works for 3 days, his helper takes over and completes the job. How long did it take him? (Draw model.)

51. Mr. Santos is 28 years older than his son Roberto. In 12 years, Mr. Santos will be twice as old as Roberto will be then. How old is each now?

52. A jet travels 500 kilometers in 40 minutes with a tail wind. Returning, the jet takes 50 minutes to cover the same distance. What is the rate of the plane and the speed of the wind?

ANSWERS TO WORD PROBLEMS—REVIEW TEST

1.   Let x = width of the rectangle

Let $3x - 10$ = length of the rectangle

Equation: $2x + 2(3x - 10) = 140$ cm

$$2x + 6x - 20 = 140$$
$$8x - 20 = 140$$
$$8x = 160$$
$$x = 20 \text{ cm in width}$$
$$(3x - 10) = 50 \text{ cm in length}$$

2. Let x = number of degrees in angle C

$2x$ = number of degrees in angle B

$6x$ = number of degrees in angle A

Equation: $9x = 180°$

$$x = 20° \text{ in angle C}$$
$$2x = 40° \text{ in angle B}$$
$$6x = 120° \text{ in angle A}$$

3. Let x = number of cm in width

$(x-10)$ = number of cm in length

Equation: $2x + 2(x-10) = 140$ cm

$$2x + 2x - 20 = 140 \text{ cm}$$
$$4x = 140 + 20$$
$$4x = 160 \text{ cm}$$
$$x = 40 \text{ cm width}$$
$$x - 10 = 30 \text{ cm length}$$

4. Let x = number of dimes

$98 - x$ = number of quarters

$10x$ = value in cents of dimes

$25(98 - x)$ = value in cents of quarters

$\$18.50 = 1,850$ value in cents

Equation: $10x + 25(98 - x) = 1,850$

$$10x + 2,450 = 1,850$$
$$-15x = 1,850 - 2,450$$
$$-15x = -600$$
$$x = 40 \text{ dimes}$$
$$98 - x = 58 \text{ quarters}$$

5.

	Age now	Age 2 yr from now	Age 2 yr ago
Frank		$x + 4$	$x + 6$
John		x	$x - 2$

Equation: $5(x+6) = 7(x-2)$

$$5x + 30 = 7x - 14$$
$$5x - 7x = -14 - 30$$
$$-2x = -44$$
$$x = 22 \text{ John's age now}$$
$$x + 4 = 26 \text{ Frank's age now}$$

6.

	Age now	Age 7 yr ago
Nancy	$x-4$	$x-11$
Jane	x	$x-7$

Equation: $x - 11 + x - 7 = 18$

$2x - 18 = 18$

$2x = 36$

$x = 18$ Jane's age now

$x - 4 = 14$ Nancy's age now

7.

$$\boxed{x} + \boxed{y} = \boxed{\begin{array}{c} 12,800 \\ \hline \text{yield } 966.00 \end{array}}$$

Equations: (1) $x + y = 12,800$

(2) $.07x + .08y = 966.00$

Move decimal point 2 places to the right.

Multiply (1) by -7 $-7x - 7y = -89,600$

$\underline{7x + 8y = 96,600}$

Add (1) and (2) $1y = \$7,000$ invested @ 8%

Substitute \$7,000 for y in equation (1)

$x + 7,000 = 12,800$

$x = \$5,800$ invested @ 7%

8.

Bonds		Pref.Stock		Yield
x	$+$	y	$=$	40,000.00
				Yield 3,620.00

Equations: (1) $\quad\quad x + y = 40,000$

(2) $.09\frac{1}{2}x + .08y = 3620.00$

Move decimal point 2 places to the right.

$$9\frac{1}{2}x + 8y = 362,000$$

(Multiply (1) by -8

$$-8x - 8y = -320,000$$
$$9\frac{1}{2}x + 8y = 362,000$$

Add (1) and (2) $\quad\quad 1\frac{1}{2}x = 42,000$

$$\frac{3}{2}x = 42,000$$

$$x = \$28,000 \text{ invested in } 9\frac{1}{2}\% \text{ bonds}$$

Substitute \$28,000 for x in (1)

$$28,000 + y = 40,000$$
$$y = \$12,000 \text{ invested in } 8\% \text{ preferred stock}$$

9.

Equations: (1) $x + y = 30$

　　　　　　　(2) $.35x + .65y = .55(30)$

Move decimal point 2 places to the right.

$$35x + 65y = 1,650$$

Multiply (1) by -35 $-35x - 35y = -1,050$

$$\underline{35x + 65y = 1,650}$$

Add (1) and (2) $30y = 600$

　　　　　　　　　　　　　$y = 20$ oz of 65%

Substitute 20 for y in (1) $x + 20 = 30$

　　　　　　　　　　　　　$x = 10$ oz of 35%

10.

	Time to do job	Part done in 1 hr
Plumber	6 hr	$\frac{1}{6}$
Helper	9 hr	$\frac{1}{9}$
Together	x hr	$\frac{1}{x}$

Equation: $\dfrac{1}{6} + \dfrac{1}{9} = \dfrac{1}{x}$ LCD $= 18x$ $3x + 2x = 18$

　　　　　　　　　　　　　　　　　　　　　$5x = 18$

　　　　　　　　　　　　　　　　　$x = 3\dfrac{3}{5}$ hr together

11.

	Hours to finish job	*Part done in 1 hr*
Expert	6	$\frac{1}{6}$
Secretary	7	$\frac{1}{7}$
Together	x	$\frac{1}{x}$

Part of job completed $\frac{1}{3}$ by expert

Equation: $\dfrac{1}{7} = \dfrac{2}{3x}$

LCD $= 21x$ $3x = 14$

$x = 4\frac{2}{3}$ hr for secretary to complete job

12.

Move decimal point 2 places to the right.

$$225 + 0x = 180 + 12x$$
$$45 = 12x$$
$$x = 3\frac{3}{4} \text{ quarts}$$

13.

	Hours to fill tank	Part filled in 1 hr
1st pipe	10 hr	$\frac{1}{10}$
2nd pipe	12 hr	$\frac{1}{12}$
3rd pipe	15 hr (drain)	$-\frac{1}{15}$ drained
Together	x	$\frac{1}{x}$

Equation: $\dfrac{1}{10}+\dfrac{1}{12}+\dfrac{-1}{15}=\dfrac{1}{x}$

$\text{LCD} = 60x$ $6x + 5x - 4x = 60$

$$7x = 60$$

$$x = 8\tfrac{4}{7} \text{ hr to fill tank}$$

14.

	Hours to complete job	Part completed in 1 hr
Typist	6 hr	$\frac{1}{6}$
Assistant	9 hr	$\frac{1}{9}$
Together	x hr	$\frac{1}{x}$

Part completed $\frac{3}{6}+\frac{3}{9}=\frac{5}{6}$ of job completed

Equation: $\left(\frac{1}{2},5\right).$

$\text{LCD} = 6x$ $\dfrac{\cancel{6}x(1)}{\cancel{6}} = \dfrac{\cancel{6}\cancel{x}(1)}{\cancel{6}\cancel{x}}$

$$x = 1 \text{ hr for experienced typist to finish job}$$

15.

	Distance	Rate	Time
Shorter route	x	50	$x/50$
Longer route	$x + 20$	55	$x+20/55$

Equation: $\dfrac{x}{50} = \dfrac{x+20}{55}$

LCD = 550 $11x = 10x + 200$

$x = 200$ miles for shorter route

$x + 20 = 220$ miles for longer route

16.

	Distance	Rate	Time
1st motorcyclist	$52x$	52	x
2nd motorcyclist	$54x$	54	x

Equation: $52x + 54x = 318$

$106x = 318$

$x = 3$ hours

17.

$140 \cdot x = 130 \cdot 14$

$140x = 1{,}820$

$x = 13$ inches (distance from fulcrum that 140-pound weight must be placed)

18.

	Distance	Rate	Time
Slow tuna boat	$24x$	x	24
Faster tuna boat	$24(x + 5)$	$x + 5$	24

Equation:

$$24x + 24(x + 5) = 1,080 \text{ miles}$$
$$24x + 24x + 120 = 1,080$$
$$48x + 120 = 1,080$$
$$48x = 1,080 - 120$$
$$48x = 960$$
$$x = 20 \text{ miles per hour for slow tuna boat}$$
$$x + 5 = 25 \text{ miles per hour for faster tuna boat}$$

19.

Equations: (1) $x + y = 5,000$

(2) $1.50x + 1.00y = 6,500.00$

Move decimal point 2 places to the right.

$$150x + 100y = 650,000$$

Multiply (1) by -100 $-100x - 100y = -500,000$

$$\underline{150x + 100y = 650,000}$$

Add (1) and (2) $50x = 150,000$

$$x = 3,000 \text{ adults}$$

Substitute 3,000 for x in (1) $3,000 + y = 5,000$

$$y = 2,000 \text{ children}$$

20. Let x = cost of 1 bricklayer's work per day

y = cost of 1 apprentice's work per day

(Multiply by -3) $8x + 4y = \$1,280$

(Multiply by 2) $10x + 6y = \$1,680$

$$-24x - 12y = -3,840$$
$$20x + 12y = 3,360$$
$$\overline{\hspace{3cm}}$$
$$-4x = -480$$

$x = \$120$ cost per day for bricklayer

$y = \$80$ cost per day for the apprentice

21.

	Distance	Rate	Time
Going	480	$r - x$	$1\frac{1}{2}$
Returning	480	$r + x$	1

Equation 1 (Proportion 1)   Equation 2 (Proportion 2)

$$\frac{480}{r-x} = \frac{3}{2} \qquad\qquad \frac{480}{r+x} = \frac{1}{1}$$

(Product of means equals product of extremes)

Equation 1 $3r - 3x = 960$

Equation 2 $r + x = 480$

(Multiply by 3) $3r - 3x = 960$

$$3r + 3x = 1,440$$
$$\overline{\hspace{3cm}}$$
$$6r = 2,400$$
$$r = 400 \text{ mph}$$
$$x = 80 \text{ mph speed of tailwind}$$

22.   Let x = length of the second side
$x + 10$ = length of the first side
$2(x + 10) - 5$ = length of the third side

Equation: $x + x + 10 + 2(x + 10) - 5 = 173$
$x + x + 10 + 2x + 20 - 5 = 173$
$4x + 25 = 173$
$4x = 173 - 25$
$4x = 148$
$x = 37$ in. length of the second side
$x + 10 = 47$ in. length of the first side
$2(x + 10) - 5 = 89$ in. length of the third side

23.

	Age now	*Age in 10 years*
Suzanne	$\frac{1}{6}x$	$\frac{1}{6}x + 10$
Mother	x	$x + 10$

Equation: $\dfrac{1}{6x} + 10 = \dfrac{3}{8}(x + 10)$

LCD = 24

$4x + 240 = 9x + 90$
$4x - 9x = 90 - 240$
$-5x = -150$
$x = 30$ years (Mother's age now)
$\frac{1}{6}x = 5$ years (Suzanne's age now)

24.

Wrestler 1   Wrestler 2

$$\boxed{\begin{array}{c} x \\ \text{lb} \end{array}} \;+\; \boxed{\begin{array}{c} y \\ \text{lb} \end{array}} \;=\; \boxed{\begin{array}{c} 450 \\ \text{lb} \end{array}}$$

.15x gain + -10y loss = 461.25 lb

$$100\% + 15\% \text{ gain} = 115\%$$
$$100\% - 10\% \text{ loss} = 90\%$$
$$1.15x + .90y = 461.25$$

Move decimal point 2 places to the right.

$$115x + 90y = 46,125 \text{ new weight}$$
(Multiply by -90) $$x + y = 450 \text{ original weight}$$

$$115x + 90y = 46,125$$
$$\underline{-90x - 90y = -40,500}$$
$$25x = 5,625$$
$$x = 225 \text{ lb original weight of Wrestler 1}$$
$$x + y = 450$$
$$225 + y = 450$$
$$y = 225 \text{ lb original weight of Wrestler 2}$$

25.   Let x = number of oil companies presently

$\frac{1}{5}x - 2$ = increase of oil companies

Equation:

$$x + \frac{1}{5}x - 2 = 22$$
LCD = 5 $$5x + 1x - 10 = 110$$
$$6x - 10 = 110$$
$$6x = 120$$
$$x = 20 \text{ oil companies at present}$$
$$\frac{1}{5} \cdot 20 - 2 = \text{increase}$$
$$4 - 2 = 2 \text{ increase}$$

a) 20 oil companies presently

b) 2·42 million = 84 million gallons of gasoline

c) 20·42 million equals 840 million gallons of gasoline

26.

Equations: (1) $\qquad x + y = 300$ lb

(2) $2.\underline{10}x + 1.\underline{90}y = 2.\underline{00}(300)$

Move decimal point 2 places to the right.

$$210x + 190y = 60,000$$

Multiply (1) by -190

$$-190x - 190y = 57,000$$
$$\underline{210x + 190y = 60,000}$$

Add (1) and (2) $\qquad 20x = 3,000$

$$x = 150 \text{ pounds @ \$2.10 per pound}$$

Substitute 150 for x in (1)

$$150 + y = 300$$

$$y = 150 \text{ pounds @ \$1.90 per pound}$$

27.

$$\boxed{\begin{array}{c}\text{Stocks}\\ x\end{array}} + \boxed{\begin{array}{c}\text{Bonds}\\ y\end{array}} = \boxed{\begin{array}{c}40,000\\ \hline \text{yield}\\ 2,720.00\end{array}}$$

6%   7%

.06   .07

Equations:
(1) $x + y = \$40,000$
(2) $.06x + .07y = \$2,720.00$

Move decimal point 2 places to the right.

$$6x + 7y = 272,000$$

Multiply (1) by -6

$$-6x - 6y = -240,000$$

Add (1) and (2) $y = \$32,000$ invested in bonds at 7%

Substitute 32,000 for y in (1)

$$x + 32,000 = 40,000$$
$$x = \$8,000 \text{ invested in stocks at } 6\%$$

28.

$$\boxed{ x } + \boxed{ y } = \boxed{\begin{array}{c}40,000 \text{ invested}\\ 2,130.00 \text{ yield}\end{array}}$$

$4\frac{1}{2}\%$ 6%

$.04\frac{1}{2}$ .06

Move decimal point 2 places to the right.

ARCO ■ *Step-by-Step: How to Solve Algebra Word Problems*

Equations:

$$(1) \qquad x + y = \$40,000$$

$$(2) \qquad 4\tfrac{1}{2} + 6y = \$213,000$$

Multiply (1) by -6 $\qquad -6x - 6y = -240,000$

$$4\tfrac{1}{2} + 6y = 213,000$$

Add (1) and (2) $\qquad -1\tfrac{1}{2}x = -27,000$

$$x = \$18,000 \text{ invested @ } 4\tfrac{1}{2}\%$$

Substitute $18,000$ for x in (1)

$$18,000 + y = 40,000$$

$$y = \$22,000 \text{ invested @ } 6\%$$

29.

	Distance	Rate	Time
1st cyclist	$30x$	30	x
2nd cyclist	$35(x-2)$	35	$(x-2)$

Equation: $\quad 30x = 35(x-2)$

$$30x = 35x - 70$$

$$30x - 35x = -70$$

$$-5x = -70$$

$$x = 14$$

$x - 2 = 12$ hr for second cyclist to overtake first cyclist

30.

	Distance	Rate	Time
1st bike (north)	$50x$	50	x
2nd bike (south)	$55x$	55	x

Equation: $\quad 50x + 55x = 420$

$$105x = 420$$

$$x = 4 \text{ hours to be 420 miles apart}$$

31.

	Nautical miles	Knots per hr	Hours
Slow boat	$24x$	x	24
Faster boat	$24(x + 8)$	$x + 8$	24

Equation:
$$24x + 24(x+8) = 1,920$$
$$24x + 24x + 192 = 1,920$$
$$48x = 1,920 - 192$$
$$48x = 1,728$$
$$x = 36 \text{ knots per hr slower boat}$$
$$x + 8 = 44 \text{ knots per hr faster boat}$$

32.

	Distance	Rate	Time
1st walker	$2\frac{1}{2}x$	$2\frac{1}{2}$	x
2nd walker	$3x$	3	x

Equation:
$$2\frac{1}{2}x + 3x = 27.5$$
$$5.5x = 27.5$$
$$55x = 275$$
$$x = 5 \text{ hours}$$
$$2\frac{1}{2} \cdot 5 = 12\frac{1}{2} \text{ miles for 1st walker}$$
$$3 \cdot 5 = 15 \text{ miles for 2nd walker}$$

33.

	Miles	Rate	Hours
Rocket	$2x$	$2x$	1
Plane	x	x	1

Equation:
$$x + 2x = 4,680$$
$$3x = 4,680$$
$$x = 1,560 \text{ mph for plane}$$
$$2x = 3,120 \text{ mph for rocket}$$

34.

	Days to complete job	Work done in 1 day
Mason	3	$\frac{1}{3}$
Helper	6	$\frac{1}{6}$
Together	x	$\frac{1}{x}$

Equation: $\dfrac{1}{3}+\dfrac{1}{6}=\dfrac{1}{x}$

LCD = $6x$ $2x + x = 6$

$x = 2$ days to complete job working together

35.

	Hours to complete job	Part done in 1 hr
John	3	$\frac{1}{3}$
Bill	x	$\frac{1}{x}$
Together	2	$\frac{1}{2}$

Equation: $\dfrac{1}{3}+\dfrac{1}{x}=\dfrac{1}{2}$

LCD = $6x$ $2x + 6 = 3x$

$x = 6$ hr for Bill to do the job alone

36.

	Hours to fill tank	Part of tank filled in 1 hr
Large pipe	10	$\frac{1}{10}$
Drain pipe	15	$-\frac{1}{15}$
Together	x	$\frac{1}{x}$

Equation: $\dfrac{1}{10}-\dfrac{1}{15}=\dfrac{1}{x}$

LCD = $30x$ $3x - 2x = 30$

$x = 30$ hours to fill pool

37.

	Days to paint house	*Part painted in 1 day*
Ralph	4	$\frac{1}{4}$
Walter	6	$\frac{1}{6}$
Together	x	$\frac{1}{x}$
Part to be completed by Walter		$\frac{3}{4}$

Equation: $\dfrac{1}{6} = \dfrac{3}{4x}$

LCD $= 12x$

$$2x = 9$$

$$x = 4\frac{1}{2} \text{ days for Walter to finish the house}$$

38.

	Days to fill reservoir	*Part filled in 1 day*
Storage pipe	15	$\frac{1}{15}$
Smaller pipe	20	$\frac{1}{20}$
Together	x	$\frac{1}{x}$
Part to be completed by smaller pipe		$\frac{13}{15}$

Equation: $\dfrac{1}{20} = \dfrac{13}{15x}$

LCD $= 60x$

$$3x = 52$$

$$x = 17\frac{1}{3} \text{ days for smaller pipe to finish}$$
$$\text{filling the reservoir}$$

39.   Let x = present office space

$1\frac{1}{2}x$ = increase in office space

$$x + 1\frac{1}{2}x = 500,000$$

$$2\frac{1}{2}x = 500,000$$

$$x = 200,000 \text{ sq ft presently}$$

$$1\frac{1}{2}x = 300,000 \text{ sq ft to be added}$$

40. Let x = cars sold in 1979

$\quad\quad$ y = decline in car sales in 1982

Set up two equations:

$\quad$ (1) $\quad\quad\quad\quad$ $y = \frac{x}{6} - 3,000$

$\quad$ (2) $\quad\quad\quad\quad$ $x - y = 300,000$

Change (1) to standard form:

$$y = \frac{x}{6} - 3,000$$
$$6y = x - 18,000$$
$$-x + 6y = -18,000$$

Add (1) and (2):

$$-x + 6y = -18,000$$
$$\underline{\quad x - y = 300,000\quad}$$
$$5y = 282,000$$
$$y = 56,400 \text{ decline in car sales in 1982}$$

Substitute 56,400 for y in (2)

$$x - 56,400 = 300,000$$
$$x = 300,000 + 56,400$$
$$x = 356,400 \text{ cars sold in 1979}$$

41. $\quad\quad$ Let x = number of cars presently on the road in L.A.

$\dfrac{1x}{5} - 10,000$ = number of cars that should be reduced

$$x - \left(\frac{1x}{5} - 10,000\right) = 480,000$$

LCD = 5

$$5x - (1x - 50.000) = 2,400,000$$
$$5x - 1x + 50,000 = 2,400,000$$
$$4x = 2,400,000 - 50,000$$
$$4x = 2,350,000$$
$$x = 587,500 \text{ cars now on the road}$$
$$\underline{-480,000 \text{ should be on the road}}$$
$$107,500 \text{ cars to be reduced}$$

42.

x | rectangle |
$x + 24$

$$\text{Let } x = \text{width}$$
$$x + 24 = \text{length}$$
$$\text{Perimeter} = 2x + 2(x + 24)$$
$$\text{Perimeter} = 2x + 2x + 48$$
$$212 = 2x + 2x + 48$$
$$212 = 4x + 48$$
$$212 - 48 = 4x$$
$$164 = 4x$$
$$x = 41 \text{ ft width}$$
$$x + 24 = 65 \text{ ft length}$$

43.

The radius of the larger circle is 14 ft and the radius of the smaller circle is 7 ft. Using the formula $A = \pi r^2$ and $\pi = \frac{22}{7}$, we have

$$A = \frac{22}{\cancel{7}} \cdot \cancel{49}^{\,7} = 154 \text{ sq ft}$$

$A = 154$ sq ft for the smaller circle

$A = \pi r^2$

$$A = \frac{22}{\cancel{7}} \cdot \frac{\cancel{196}^{\,28}}{1} = 616 \text{ sq ft for larger circle}$$

616 sq ft -154 sq ft $= 462$ sq (area of shaded part
    of the two concentric circles)

44. r = rate of boat in still water

x = rate of current

↓ Downstream

$d = 50$ miles

$r = r + x$

$t = 2\frac{1}{2}$ hr

Equation 1

$\dfrac{50}{r+x} = 2\frac{1}{2}$ hr

L.C.D. $= 2(r+x)$

$100 = 5r + 5x$

↑ Upstream

$d = 50$ miles

$r = r - x$

$t = 4$ hr

Equation 2

$\dfrac{50}{r-x} = 4$ hr

L.C.D. $= (r-x)$

$50 = 4r - 4x$

	(1)	$5r + 5x = 100$
	(2)	$4r - 4x = 50$
Multiply (1) by 4		$20r + 20x = 40$
Multiply (2) by 5		$20r - 20x = 250$
Add (1) and (2)		$40r = 650$

$$r = 16\frac{1}{4} \text{ mph (rate of boat in still water)}$$

Substitute $16\frac{1}{4}$ for r in (1)

$$4\left(16\frac{1}{4}\right) - 4x = 50$$

$$x = 3\frac{3}{4} \text{ mph (rate of current)}$$

45.

	Distance	Rate	Hours
Federal bus	$56x$	56 mph	x
2nd bus	$60\left(x - \frac{1}{2}\right)$	60 mph	$\left(x - \frac{1}{2}\right)$

Equation:

$$56x = 60\left(x - \frac{1}{2}\right)$$
$$56x = 60x - 30$$
$$-4x = -30$$
$$x = 7\frac{1}{2} \text{ hrs travel time for Federal Railways bus}$$
$$\left(x - \frac{1}{2}\right) = 7 \text{ hrs for second bus to overtake Federal Railways bus}$$

46.

$$I - II + III = IV$$
$$40\%(20) - 40\%(x) + 100\% \cdot x = 50\% \cdot (20)$$
$$.40(20) - .40x + 1.00x = .50(20)$$

Move decimal point 2 places to the right.

$$40(20) - 40(x) + 100x = 50(20)$$
$$800 - 40x + 100x = 1,000$$
$$60x = 1,000 - 800$$
$$60x = 200$$
$$x = 3\frac{1}{3} \text{ quarts}$$

$3\frac{1}{3}$ quarts must be drained off and added

47. Let x = age of older man

Let y = age of younger man

Equations: (1) $x + y = 62$

(2) $\dfrac{y+6}{2} = \dfrac{x+2}{3}$

Convert (2) to standard form

$$3y + 18 = 2x + 4$$
$$-2x + 3y = -14$$

Multiply (1) by 2 $2x + 2y = 124$

Add (1) and (2) $\underline{-2x + 3y = -14}$

$$5y = 110$$

$y = 22$ age of younger man

Substitute 22 for y in (1) $x + 22 = 62$

$x = 40$ age of older man

48. $6n° = 180°$

$n° = 30°$

Angle C $= 30°$

Angle B $= 60°$

Angle A $= 90°$

49.

	Hours to empty tank	Part emptied in 1 hr
Old pump	8	$\tfrac{1}{8}$
New pump	4	$\tfrac{1}{4}$
Hours for new pump	x	$\tfrac{1}{x}$
What part completed	$\tfrac{3}{8}$	Needs to be done $\tfrac{5}{8}$

Equation: $\dfrac{1}{4} = \dfrac{5}{8x}$

LCD $= 8x$ $8x = 20$

$x = 2\tfrac{1}{2}$ hours for new pump to complete the job

50.

	Days to build wall	Work done in 1 day
Bricklayer	8	$\frac{1}{8}$
Helper	10	$\frac{1}{10}$
Days for helper to finish job	x	$\frac{1}{x}$
Part completed	$\frac{3}{8}$	To be completed $\frac{5}{8}$

Equation: $\dfrac{1}{10} = \dfrac{5}{8x}$

LCD $= 40x$

$$4x = 25$$

$$x = 6\tfrac{1}{4} \text{ days for helper to finish building the wall}$$

51.

	Age now	Age in 12 years
Mr. Santos	$x + 28$	$x + 40$
Roberto	x	$x + 12$

Equation: $x + 40 = 2(x + 12)$

$$x + 40 = 2x + 24$$

$$-x = -16$$

$$x = 16 \text{ age of Roberto now}$$

$$x + 28 = 44 \text{ age of Mr. Santos now}$$

52.

	Distance	Rate	Time
With wind	500	$r + x$	$\frac{2}{3}$
Against wind	500	$r - x$	$\frac{5}{6}$

Equations:

(1) $\dfrac{500}{r + x} = \dfrac{2}{3}$ (2) $\dfrac{500}{r - x} = \dfrac{5}{6}$

$\quad\quad 2r + 2x = 1,500 \quad\quad\quad 5r - 5x = 3,000$

Multiply (1) by 5 $10r + 10x = 7,500$

Multiply (2) by 2 $10r - 10x = 6,000$

Add (1) and (2) $\quad 20r = 13,500$

$$r = 675 \text{ km per hr speed of plane}$$

$$75 \text{ km per hr speed of wind}$$

Chapter Seventeen

Problems Involving Quadratic Equations

Quadratic equations are equations of the second degree, that is, equations that have an exponent of 2.

EXAMPLES:

a. $x^2 - 10x + 25 = 0$

b. $3x^2 - 8x + 5 = 0$

c. $15x^2 - 15x - 30 = 0$

This chapter will focus on the solution of quadratic equations of *3 terms* by use of the *factoring method*.

Factoring is the reverse process of multiplication. By using the factoring method, each of the above equations will have 2 values for the variable.

a. $(+5, +5)$

b. $\left(1\tfrac{2}{3}, 1\right)$

c. $(+2, -1)$

RULE

If the product of 2 factors is zero, then one of the 2 factors is equal to zero.

EXAMPLES:

a. If $a \cdot 5 = 0$
Then $a = 0$

b.   If $(a+2)\cdot(a-5)=0$

Then $a+2=0$

$$a=-2 \;\rightarrow -2$$

Then $(a-5)=0$ $\Big\}$ 2 roots

$$a=+5 \rightarrow +5$$

c. $a(a-6)=0$

$$a=0$$

$$a-6=0$$

$$a=+6 \quad \text{The 2 roots are } (0,6).$$

d. $(2m-7)\cdot(3m+4)=0$

$$2m-7=0$$

$$2m=7$$

$$m=3\,\tfrac{1}{2}$$

$$3m+4=0$$

$$m=-1\,\tfrac{1}{3}$$

The 2 roots are $\left(3\,\tfrac{1}{2}-1\,\tfrac{1}{3}\right)$.

SOLVING QUADRATIC TRINOMIALS BY THE FACTORING METHOD

$(x + 5)$ and $(x + 4)$ are the factors of the trinomial $x^2 + 9x + 20$.

$(a - 7)$ and $(a + 5)$ are the factors of $a^2 - 2a - 35$.

$(2m + 7)$ and $(m - 4)$ are the factors of $2m^2 - 1m - 28$.

EXAMPLE 1:

The procedure for finding the two factors of a quadratic trinomial $x^2 - 8x + 15 = 0$ is as follows:

$$x^2 - 7x - 18 = 0$$

Step 1 → The 2 factors of 18 are

6	9	15
3	2	1

The 2 factors of $x^2 - 8x + 15$ are $(x - 5)$ and $(x - 3)$.

EXAMPLE 2:

Using the factoring method to find the 2 roots of the quadratic equation $x^2 - 7x - 18 = 0$.

$$x^2 - 7x - 18 = 0$$

Step 2 → Since the sign of the third term is negative, we must subtract the smaller factor from the larger to get the numerical co-efficient of the middle term which is 7.

Step 3 →
$$\begin{array}{r} -9 \\ +2 \\ \hline -7 \end{array}$$

Step 4 → When the sign of the third term is negative, the sign of the middle term will go to the *higher* of the 2 factors.

The factors of $x^2 - 7x - 18$ are $(x - 9)$ and $(x + 2)$.

The two roots are $(9, -2)$.

RULE I

If the sign of the third term is positive, the sign of the middle term goes to *both* factors of the third term.

RULE II

If the sign of the third term is negative, the sign of the middle term goes to the *higher* factor of the third term.

Now we will use the rules for finding the factors of a quadratic equation and the 2 roots.

EXAMPLE 3:

$$m^2 - 10m + 24 = 0$$

The 2 factors of $m^2 - 10m + 24 = 0$

$$\text{are} (m - 6)(m - 4) = 0$$

$$m - 6 = 0$$

$$m = +6 \leftarrow$$

$$m - 4 = 0 \quad \Big\} 2 \text{ roots}$$

$$m = +4 \leftarrow$$

The 2 roots are (+6, +4).

PRACTICE PROBLEMS: FINDING ROOTS I

Find the 2 roots of the following equations by using the factoring method.

1. $a^2 - 7a + 10 = 0$

2. $r^2 + 12r + 20 = 6$

3. $r^2 - 1r - 25 = 0$

Check your answers with the solutions that follow.

ANSWERS TO PRACTICE PROBLEMS: FINDING ROOTS I

1. $(a - 5)(a - 2) = 0$

$$a - 5 = 0 \qquad a - 2 = 0$$
$$a = 5 \qquad a = 2$$

The 2 roots are (5, 2).

2. $(r + 10)(r + 2) = 0$

$$r + 10 = 0 \qquad r + 2 = 0$$
$$r = -10 \qquad r = -2$$

The 2 roots are (–10, –2).

3. $(r - 8)(r + 7) = 0$

$$r - 8 = 0 \qquad r + 7 = 0$$
$$r = 8 \qquad r = -7$$

The 2 roots are (8, –7).

When the numerical coefficient of the x^2 term is greater than 1, we can use the *cross product method* to solve for the 2 roots.

EXAMPLE 1:

$3x^2 - 8x + 5 = 0$

$$(3x \quad -5)$$
$$(1x \quad -1)$$

$3x \cdot -1 = -3x$ cross product

$1x \cdot -5 = -5x$ cross product

The two factors are $(3x - 5)(1x - 1) = 0$

$$3x - 5 = 0 \qquad 1x - 1 = 0$$
$$x = 1\tfrac{2}{3} \qquad 1x = 1$$

The two roots are $\left(1\tfrac{2}{3}, 1\right)$.

EXAMPLE 2:

$2a^2 - 5a - 18 = 0$. Find the 2 roots.

$$2a^2 - 5a - 18 = 0$$

The 2 factors of $2a^2$ are ←——Step 2 | | Step 1 ——→ The 2 factors of 18 are

$2a$ 9   6   18
$1a$ 2   3   1

Step 3 ——→ The sign of the third
term is negative $(-)$.
Therefore, we subtract
one of the cross products
from the other to get $5a$.

Sign of the middle term

$(2a \quad -9)$ $1a \cdot -9 = -9a$ cross product
$(1a \quad +2)$ $2a \cdot 2 = \underline{\quad 4a}$ cross product
$\qquad\qquad\qquad\qquad -5a$

Step 4   Since the sign of the third term is negative, then the sign of the middle
term will go to the higher of the two cross products.

That is,
$$-9a$$
$$\underline{+4a}$$
$$-5a$$

The 2 factors are $(2a - 9)(1a + 2) = 0$

$$2a - 9 = 0 \qquad 1a + 2 = 0$$
$$2a = 9 \qquad\qquad a = -2$$
$$a = 4\tfrac{1}{2}$$

The 2 roots are $\left(4\tfrac{1}{2}, -2\right)$.

PRACTICE PROBLEMS: FINDING ROOTS II

Use the factoring method to find the 2 roots of the following equations:

1. $2a^2 - 11a + 5 = 0 = 0$

2. $3m^2 - 2m - 8 = 0$

3. $7x^2 - 10x - 8 = 0$

Check your answers with the solutions that follow.

ANSWERS TO PRACTICE PROBLEMS: FINDING ROOTS II

1. $(2a - 1)(a - 5) = 0$

$$2a - 5 = 0 \qquad a - 5 = 0$$
$$2a = 1 \qquad\quad a = +5$$
$$a = \frac{1}{2}$$

The 2 roots are $\left(\frac{1}{2}, 5\right)$.

2. $(3m + 4)(m - 2) = 0$

$$3m + 4 = 0 \qquad m - 2 = 0$$
$$3m = -4 \qquad\quad m = +2$$
$$m = -1\frac{1}{3}$$

The 2 roots are $\left(-1\frac{1}{3}, 2\right)$.

3. $(7x + 4)(x - 2) = 0$

$$7x + 4 = 0 \qquad x - 2 = 0$$
$$7x = -4 \qquad\quad x = +2$$
$$x = -\frac{4}{7}$$

The 2 roots are $\left(-\frac{4}{7}, 2\right)$.

WORD PROBLEMS INVOLVING QUADRATIC EQUATIONS

STUDY PROBLEM 1

$$x + 5$$

If the length of the rectangle is 5 feet longer than the width and the area is 50 square feet, what are the dimensions of the rectangle? Area = length · width.

$$\text{Let } x = \text{ width}$$
$$x + 5 = \text{ length}$$
$$x(x + 5) = \text{area}$$

Equation:
$$x(x + 5) = 50 \text{ sq ft}$$
$$x^2 + 5x - 50 = 0$$
$$(x + 10)(x - 5) = 0$$
$$x + 10 = 0$$
$$x = -10$$
$$x - 5 = 0$$
$$x = +5 \text{ ft}$$

Roots are (5 ft, −10)

However, the length of a line is never negative, so in this instance, the width is 5 feet and the length is 10 feet.

STUDY PROBLEM 2

One number is 5 times a smaller number. If the product of their squares is 2,600, what are the numbers?

Let $x =$ smaller number

$5x =$ larger number

$x^2 =$ square of smaller number

$25x^2 =$ square of larger number

$$x^2 + 25x^2 = 2,600$$
$$26x^2 = 2,600$$
$$x^2 = 100$$
$$\sqrt{x^2} = \sqrt{100}$$
$$x = +10 \text{ and } = -10$$
$$5x = +50 \text{ and } 5x = -50$$

STUDY PROBLEM 3

Solve the following quadratic equation:

$$\frac{1}{x} + \frac{1}{x+2} = \frac{5}{12}$$

LCD $= 12(x)(x + 2)$

Clear fractions: $12(\cancel{x})(x+2)\dfrac{1}{\cancel{x}} + \dfrac{1}{\cancel{x+2}}12(x)(\cancel{x+2}) = \dfrac{5}{\cancel{12}}\cancel{12} \cdot x(x+2)$

$$12(x+2)+12x = 5x(x+2)$$
$$12x+24+12x = 5x^2 +10x$$
$$0 = 5x^2 -14x-24$$
$$0 = (5x+6)(x-4)$$
$$0 = 5x+6 \qquad 0 = x-4$$
$$-6 = 5x \qquad\qquad 4 = x$$
$$-1\tfrac{1}{5} = x$$

Roots are $\left(-1\tfrac{1}{5}, 4\right)$

STUDY PROBLEM 4

If 48 square inches are subtracted from the area of a square, the result would be equal to twice the perimeter. What is the length of the side of the square?

$$\text{Area of square} = x^2$$
$$\text{Perimeter} = 4x$$
$$x^2 - 48 = 2(4x)$$
$$x^2 - 48 = 8x$$
$$x^2 - 8x - 48 = 0$$
$$(x-12)(x+4) = 0$$
$$x = +12 \ \left(\text{length of side of square}\right)$$
$$x = -4$$

STUDY PROBLEM 5

The altitude of the triangle is 12 in. longer than the base. The area of the triangle is $A = \frac{1}{2}$ base · altitude.

Let x = base of triangle
$x + 12$ = altitude of triangle

Find the base and the altitude if the area is equal to 32 square inches.

$$A = \frac{1}{2}(x)\cdot(x+12)$$
$$32 = \frac{1}{2}(x)\cdot(x+12)$$
$$64 = x(x+12)$$
$$64 = x^2 + 12x$$
$$0 = x^2 + 12x - 64$$
$$0 = (x+16)(x-4)$$
$$x = +4 \text{ in.base of the triangle}$$
$$x + 12 = 16 \text{ in. altitude of the triangle}$$

STUDY PROBLEM 6

The area of a rectangle is equal to the area of a square. The length of the rectangle is 6 inches less than twice the side of the square. The width of the rectangle is 4 inches less than the side of the square. What is the length of one side of the square?

$$x^2 = L \cdot W$$
$$L = (2x-6)$$
$$W = (x-4)$$

Equation: $(2x-6)(x-4) = x^2$
$$2x^2 - 14x + 24 = x^2$$
$$2x^2 - x^2 - 14x + 24 = 0$$
$$x^2 - 14x + 24 = 0$$
$$(x-12)(x-2) = 0$$
$$x = +12 \text{ in. length of side square}$$
$$x = +2$$

STUDY PROBLEM 7

The perimeter of a rectangle is 52 centimeters and the area is 144 square centimeters. What are the dimensions of the rectangle?

$$\text{Let } x = \text{ width}$$
$$y = \text{length}$$
$$2x + 2y = \text{ perimeter}$$
$$2x + 2y = 52$$

Simplifying by dividing both members by 2, we get factoring.

$$x + y = 26$$
$$x = 26 - y$$
$$A = L \cdot W \text{ area of rectangle}$$
$$144 = (26 - y) \cdot y$$
$$144 = 26y - y^2$$
$$y^2 - 26y + 144 = 0$$
$$(y - 18)(y - 8) = 0$$
$$y - 18 = 0$$
$$y = +18 \text{ length}$$
$$y - 8 = 0$$
$$x = +8 \text{ width}$$

PRACTICE PROBLEMS: QUADRATIC EQUATIONS

1. If the area of the rectangle is 45 square centimeters, what is the length and the width?

x

$x + 12$

2. One number is larger than another by 15. If the product of the two numbers is 154, what are the 2 numbers?

3. The length of a rectangle is 8 feet longer than its width, which has the same dimension as the side of the square. If the area of the rectangle is equal to 160 square feet more than the area of the square, what are the length and width of the rectangle? (Draw diagram.)

4. One holiday, Mr. Anderson drove his family to a picnic ground 60 miles from his house. Returning home that evening, he increased his average speed 9 miles per hour above the rate on the way to the picnic. This increase in miles per hour reduced his return time by 20 minutes. What was his average speed going and returning? (Use model for distance, rate and time.)

5. Working together, an old pump and a new pump can fill a tank in 2 hours. If the new pump can fill the tank in 3 hours less time than the old pump, how many hours will it take each pump to fill the tank? (Use work problem model.)

6. The sum of two numbers is 25 and the difference of their squares is 75. What are the 2 numbers?

7. The perimeter of rectangle ABCD is equal to 34 rods. If the area of the rectangle is 60 square rods, what are the dimensions of the rectangle?

Check your answers with the solutions that follow.

ANSWERS TO PRACTICE PROBLEMS: QUADRATIC EQUATIONS

1. Let $x =$ width

$(x+12) =$ length

Area $= x(x+12)$

$45 = x^2 + 12x$

$-x^2 - 12 = -45$

$x^2 + 12x - 45 = 0$

$(x+15)(x-3) = 0$ (factoring method)

$x = +3$

$x = -5$

$x = 3$ cm

$x + 12 = 15$ cm

2. Let $x =$ smaller number

$(x+15) =$ larger number

$x(x+15) = 154$

$x^2 + 15x - 154 = 0$

$(x+22)(x-7) = 0$

$x = +7$

$x = -22$

(The 2 numbers are 7 and 22 or –7 and –22.)

3.

x | square

x

x | rectangle

$x+8$

$x(x+8) =$ area of rectangle

$x^2 =$ area of square

$x(x+8) = x^2 + 160$

$x^2 + 8x = x^2 + 160$

$x^2 - x^2 + 8x = 160$

$x = 20$ ft width of rectangle

$(x+8) = 28$ ft length of rectangle

4.

	Distance	Rate	Time
Going	60	r	$60/r$
Returning	60	$r + 9$	$60/_{r+9}$

$$\frac{60}{r} - \frac{60}{r+9} = \frac{1}{3}$$

LCD $= 3r(r + 9)$

Clear fractions:

$$3\cancel{r}\,(r+9)\frac{60}{r} - 3r\,(\cancel{r+9})\frac{60}{\cancel{r+9}} = \cancel{3}r\,(r+9)\frac{1}{\cancel{3}}$$

$$180(r+9) - 180r = r(r+9)$$

$$180r + 1,620 - 180r = r^2 + 9r$$

$$-r^2 - 9r + 1,620 = 0$$

Changing signs $(-)$: $r^2 + 9r - 1,620 = 0$

$$(r+45)(r-36) = 0$$

$$r = 36 \text{ mph going}$$

$$r = -45$$

$$r + 9 = 45 \text{ mph returning}$$

5.

	Hours to fill tank	Part filled in 1 hr
Old pump	x	$1/x$
New pump	$x - 3$	$1/_{x-3}$
Together	2	$1/2$

Equation: $\dfrac{1}{x} + \dfrac{1}{x-3} = \dfrac{1}{2}$

LCD $= 2 \cdot x \cdot (x - 3)$

Clear fractions:

$$2 \cdot \cancel{x}\,(x-3)\frac{1}{\cancel{x}} + \left(\cancel{x-3}\right)\frac{1}{\cancel{x-3}} = 2 \cdot x\,(x-3)\frac{1}{2}$$

$$2(x-3)+2 \cdot x = x^2 - 3x$$

$$2x - 6 + 2x = x^2 - 3x$$

Changing signs $(-)$: $\qquad -x^2 + 7x - 6 = 0$

$$x^2 - 7x + 6 = 0$$

$$(x-6)(x-1) = 0$$

$$x = 6$$

$$x = 1$$

6 hours for old pump to fill the tank

3 hours for new pump to fill tank

6. Let $x =$ one number

$25 - x =$ other number

Equation: $\quad (25-x)^2 - x^2 = 75$

$$625 - 50x + x^2 = 75$$

$$-50x = 75 - 625$$

$$-50x = -550$$

$$x = 11 \text{ smaller number}$$

$$25 - x = 14 \text{ larger number}$$

7.

$$\text{Perimeter} = 2L + 2W$$

$$34 = 2L + 2W$$

Simplifying, divide by (2): $17 = L + W$

$$17 - W = L$$

$$\text{Area of rectangle} = L \cdot W$$

Simplifying by changing signs (?):

$$W(17 - W) = 60$$

$$-W^2 + 17W - 60 = 0$$

$$W^2 - 17W + 60 = 0$$

$$(W - 12)(W - 5) = 0$$

$W = 12 \text{ rods} : 17 - W = L, \text{ then Length} = 5 \text{ rods}$

$W = 5 \text{ rods} : 17 - W = L, \text{ then Length} = 12 \text{ rods}$

Chapter Eighteen

Trigonometric Word Problems

Trigonometry is the branch of mathematics that deals with the relations between the sides and angles of triangles.

Problems involving *right* triangles can be solved by using the *Pythagorean Theorem*. It has been proven in geometry that in any right triangle, the sum of the squares of each leg is equal to the square of the hypotenuse. This law is called the *Pythagorean Theorem*.

EXAMPLE 1:

In triangle ABC, angle C is a right angle (90°), the side opposite angle C is called the hypotenuse, and the two other sides are called the legs of the right triangle.

The Pythagorean Theorem states that THE SQUARE OF THE LENGTH OF THE HYPOTENUSE IS EQUAL TO THE SUM OF THE SQUARES OF THE LENGTH OF THE LEGS. (a) and (b) are the two legs and (c) is the hypotenuse. Therefore, the Pythagorean equation is:

$$a^2 + b^2 = c^2$$

If (a) = 3 and (b) = 4 and (c) = 5, then:

$$3^2 + 4^2 = 5^2$$
$$9 + 16 = 25$$
$$25 = 25$$

EXAMPLE 2:

In triangle ABC, angle A + angle B = 90°. If side a = 12 and side b = 9, what is the length of the hypotenuse?

If angle A + angle B = 90°, then angle C is also 90° because the sum of the angles of a triangle is 180° and triangle ABC is a right triangle. Use the Pythagorean Thorem to find the hypotenuse: $a^2 + b^2 = c^2$

(Refer to Table I at the back of this book to find square roots.)

$$12^2 + 9^2 = C^2$$
$$144 + 81 = C^2$$
$$225 = C^2$$
$$\sqrt{225} = \sqrt{C^2}$$
$$15 = C$$

Note: Although the square root of 225 is also – 15, we use the positive number because you cannot have a negative side to a triangle.

EXAMPLE 3:

One side of a right triangle is 7 feet shorter than the other. If the hypotenuse is 17 feet long, what is the perimeter of the triangle?

Let x represent the second side.

Let $x - 7$ represent the first side.

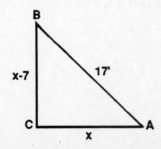

$$x^2 + (x-7) = 17^2$$
$$x^2 + x^2 - 14 + 49 = 289$$
$$2x^2 - 14x - 240 = 0$$
$$(2x - 30)(x + 8) = 0$$
$$2x - 30 = 0$$
$$2x = 30$$
$$x = 15$$
$$x + 8 = 0$$
$$x = -8$$

We reject the negative number.

The perimeter = 15 + 8 + 17 = 40 feet.

PRACTICE PROBLEMS: RIGHT TRIANGLES

1. The length of a rectangle is 5 feet longer than the width. If the diagonal is 25 feet, what are the dimensions of the sides?

2. The sides of a right triangle are consecutive even numbers. Find the length of the sides.

3. The sides of a right triangle differ by 3. Find the sides.

4. A brace wire extends from the top of a 30-foot pole to a point on the ground 38 feet from the foot of the pole. Allowing 4 feet for attaching the wire, how long is the wire?

ANSWERS TO PRACTICE PROBLEMS: RIGHT TRIANGLES

1. Let x = the width

$x + 5$ = the length

$$(x)^2 + (x+5)^2 = 25^2$$
$$x^2 + x^2 + 10x - 600 = 0$$
$$x^2 + 5x - 300 = 0$$
$$(x+20)(x-15) = 0$$
$$x = -20 \ \left(\text{reject the negative number}\right)$$
$$x = +15$$

The dimensions are: width = x = 15′

length = $x + 5 = 20′$

2. Let x = length of the smallest leg

    $x + 2$ = length of the other leg

    $x + 4$ = length of the hypotenuse

$$\text{Leg}^2 + \text{Leg}^2 = \text{Hyp}^2$$
$$x^2 + (x+2)^2 = (x+4)^2$$
$$x^2 + x^2 + 4x + 4 = x^2 + 8x + 16$$
$$x^2 + 4x + 4 = 8x + 16$$
$$x^2 + 4x + 4 - 8x - 16 = 0$$
$$x^2 - 4x + 12 = 0$$
$$(x-6)(x+2) = 0$$
$$x - 6 = 0$$
$$x = 6$$
$$x + 2 = 0$$
$$x = -2 \; (\text{reject the negative number})$$

Therefore $x = 6 \; (\text{smallest leg})$

        $x + 2 = 8 \; (\text{other leg})$

        $x + 4 = 10 \; (\text{hypotenuse})$

3. Let x = length of the smallest leg

   $x + 3$ = length of the other leg

   $x + 6$ = length of the hypotenuse

$$\text{Leg}^2 + \text{Leg}^2 = \text{Hyp}^2$$
$$x^2 + (x+3)^2 = (x+6)^2$$
$$x^2 + x^2 + 6x + 9 = x^2 + 12x + 36$$
$$x^2 + 6x + 9 - 12x = +36$$
$$x^2 - 6x + 27 = 0$$
$$(x-9)(x+3) = 0$$
$$(x-9) = 0$$
$$x = 9$$
$$x + 3 = 0$$
$$x = -3 \ \left(\text{reject the negative number}\right)$$

Therefore, $x = 9 \ \left(\text{smallest leg}\right)$

      $x + 3 = 12 \ \left(\text{other leg}\right)$

      $x + 6 = 15 \ \left(\text{hypotenuse}\right)$

4. Let x represent the length AB of the wire.

$$x^2 = (38)^2 + (30)^2$$
$$x^2 = 1,444 + 900$$
$$x^2 = 2,344$$
$$\sqrt{x^2} = \sqrt{2,344}$$
$$x = 48.4 \text{ ft}$$
$$48.4 + 4 = 52.4 \text{ length of wire}$$

TRIGONOMETRIC FUNCTIONS

To solve some problems involving right triangles, as well as other kinds of triangles, you need to be acquainted with the trigonometric functions that define the ratios between the angles and sides of the triangle. These three functions are called:

SINE

COSINE

TANGENT

An imaginative student came up with a scheme to remember the definitions of the three trigonometric functions. He devised the word: SOHCAHTOA

SOH: Sine = Opposite Leg : the Hypotenuse
CAH: Cosine = Adjacent Leg : the Hypotenuse
TOA: Tangent = Opposite Leg : the Adjacent Side

STUDY PROBLEM 1

How long must a ramp be to reach across a stream 17.2 feet wide, if the angle of elevation with the opposite bank is 22°?

In the figure, we know the angle and the adjacent leg, and wish to find the hypotenuse. We need to know, therefore, the ratio that compares both A (adjacent leg) and H (the hypotenuse). Looking at the information given, we know that the COSINE—that is, the relationship between the opposite leg and the hypotenuse—is 22°. Using the COSINE then, we can solve for the hypotenuse.

Angles can be translated into decimals. Fortunately for the student of trigonometry, the relationships between sines, cosines, and tangents have already been compiled and can be found in TABLE 2 at the end of this book.

Examples from TABLE 2

Angle	Sine	Cosine	Tangent
22°	0.375	0.927	0.404
23°	0.397	0.921	0.424

Using SOH*CAH*TOA, we obtain the equation:

$$\cos = \frac{adj.\ leg}{Hyp}$$

$$\cos = \frac{17.2}{AB} \qquad \text{(hypotenus)}$$

Using TABLE 2, substitute the corresponding decimal for 22°.

$$\frac{0.927}{1} = (1)(17.2)$$

$$AB = \frac{17.2}{0.927}$$

$$AB = 18.55$$

$$AB = 18.6 \text{ feet}$$

STUDY PROBLEM 2

How high up a wall does a 10-foot ladder reach, if it has an angle of elevation of 63°?

Draw the picture

Find the ratio that includes the hypotenuse and the opposite leg.

That ratio is SOH or $Sin = \dfrac{\text{opposite leg}}{\text{hypotenuse}}$

$$Sin\ 63° = \dfrac{x}{10}$$

Substitute the value for sin 63° from TABLE 2 and solve for x (opposite leg):

$$.891 = \dfrac{x}{10}$$
$$8.91 = x$$

STUDY PROBLEM 3

From a plane flying at an altitude of 2,000 feet, the angle of depression (the angle looking down) to an airport tower is 62°. Find, to the nearest foot, the distance the plane must fly to be directly over the tower assuming there is no change in altitude.

The angle of depression is the angle formed by the plane's path (AB) and the sightline to the tower (AC). Draw a right triangle whose legs are the plane's path and the altitude of the plane directly above the tower, and whose hypotenuse is the sight-line to the tower.

You have an angle and the opposite leg and you want to find the adjacent leg. Therefore, use:

$$\text{Tangent} = \frac{\text{opposite leg}}{\text{adjacent leg}}$$

$$\text{Tan } 62° = \frac{2000}{x}$$

Substitute value of tan 62° from Table 2 and solve for x.

$$1.881 = \frac{2000}{x}$$

$$x = \frac{2000}{1.881}$$

$$x = 1063 \text{ ft.}$$

PRACTICE PROBLEMS: TRIGONOMETRIC FUNCTIONS

1. A cable for a 15-foot pole is attached at a 60° angle to the ground. Find the length of the cable to the nearest tenth of a foot.

2. An engineer surveying a building set up his transit 100 feet from the base of the building. The angle of elevation with the top of a building was 61°. What is the height of the building?

3. At a point on the ground 50 feet from the foot of a tree, the angle of elevation to the top of the tree is 38°. How high is the tree?

4. From an airplane flying at an elevation of 4,000 feet over the ocean, the angle of depression with the coast of an island is 15°. How many miles must the airplane still go to be directly over the island?

5. From the top of a lighthouse 200 feet high, the angle of depression of a boat at sea measures 36°. Find, to the nearest foot, the distance from the boat to the foot of the lighthouse.

6. In triangle ABC, angle C is a right angle. Angle A is 23° and the hypotenuse is 44 inches. What is the length of AC?

ANSWERS TO PRACTICE PROBLEMS: TRIGONOMETRIC WORD PROBLEMS

1.

$$\text{Sin } 60° = \frac{15}{x}$$

$$.866 \cdot x = 15$$

$$x = \frac{15}{.866}$$

$$x = 17.3 \text{ ft}$$

2.

$$\text{Tan } 61° = \frac{x}{100}$$

$$1.804 = \frac{x}{100}$$

$$180.4 \text{ ft} = x$$

3.

$$\text{Tan } 38° = \frac{\text{opp. leg}}{\text{adj. leg}}$$

$$\text{Tan } 38° = \frac{x}{50}$$

$$50(.781) = x$$

$$39.05 \text{ ft} = x$$

4.

$$\text{Tan } 15° = \frac{4000}{D}$$

$$.268 = \frac{4000}{D}$$

$$D = \frac{4000}{.268}$$

$$D = 14,925 \text{ ft}$$

5.

$$\text{Tan } 36° = \frac{200}{x}$$

$$x(.727) = 200$$

$$x = \frac{200}{.727}$$

$$x = 275 \text{ ft}$$

6.

$$\text{Cos } 23° = \frac{AC}{44}$$

$$(.921)(44) = C$$

$$40.5'' = AC$$

Chapter Nineteen

Probabilities, Permutations, and Combinations Probabilities

In the previous chapters of this book, each problem had a precise answer. That answer remained the same no matter how many times the problem was presented. There are, however, many problems in which the answers are uncertain. Nonetheless, some answers are *more probable* than others.

Many industries find it useful to determine which answers or solutions are more probable than others. Some auto insurance companies, for example, base some rates on the probability that men and women under 25 years of age will have the most accidents. A gambler in Las Vegas places a bet based on the probability that the number 7 will turn up on the dice before the number 3. Engineers launch space telescopes based on a high probability that the mechanism will function perfectly. A doctor prescribes a particular medication based partially on the probability that there will be few negative side effects. Probability is fundamental to many fields, including business, physics, politics, and the military.

It is possible to set up equations to reflect probabilities.

STUDY PROBLEM 1

From a bag containing twelve white balls and 18 red ones, one ball is drawn at random.

(a) What is the probability that the ball drawn is white?

(b) What is the probability that the ball drawn is red?

> 12 = the white balls
> 18 = the red balls
> 30 = total number of balls

(a) To find the probability of drawing a white ball, divide the white balls by the total number of balls and reduce the fraction to the lowest common denominator.

$$\frac{12}{30} = \frac{2}{5}$$

Answer: There is a $\frac{2}{5}$ probability that the ball drawn is white.

(b) To find the probability of drawing a red ball, divide the red balls by the total number of balls and reduce the fraction to the lowest common denominator.

$$\frac{18}{30} = \frac{3}{5}$$

Answer: There is a $\frac{3}{5}$ probability that the ball drawn is red. It can also be said that there is a $\frac{1}{5}$ greater probability that the ball drawn will be red.

SAMPLE SPACES

Toss a coin 30 times. Does your experiment show that heads comes up $\frac{15}{30}$ or $\frac{1}{2}$ the time? If not, remember that we are only talking about probabilities, not exact outcomes. A *sample space* for this experiment is (H, T).

There are two possible *events* when you toss a coin: heads (H) and tails (T). When an event involves a single element of the sample space, it is called a *simple event*. (H) and (T) are both simple events.

Either one of the two outcomes H or T is equally likely to occur. The probability of the simple event (H), then, is the same as the probability of (T). Therefore:

$$P(H) = P(T) = \frac{1}{2}$$

Since there are two possible outcomes, one for heads and one for tails, then P = 2. The probability, then, that heads will turn up is:

$$HP \text{ or } \frac{1}{2}$$

The probability that T will turn up is:

$$TP \text{ or } \frac{1}{2}$$

Therefore, the *chances are* one in two or one-half that a head will turn up in one toss. If you toss the coin 20 times, the probability is $\frac{10}{20}$ or still one-half.

STUDY PROBLEM 2

Two dice (with numbers 1 through 6) are rolled and the numbers are recorded. Find the probability of each event below:

Event *A:* (the sum is 2 or 3)
Event *B:* (the sum is greater than 8)

Solution: First we must define the *sample spaces* (all the possible outcomes) for rolling two dice.

$$S = (1, 1) (1, 2) (1, 3) (1, 4) (1, 5) (1, 6) (2, 1) (2, 2) (2, 3) (2, 4) (2, 5) (2, 6)$$
$$(3, 1) (3, 2) (3, 3) (3, 4) (3, 5) (3, 6) (4, 1) (4, 2) (4, 3) (4, 4) (4, 5) (4, 6)$$
$$(5, 1) (5, 2) (5, 3) (5, 4) (5, 5) (5, 6) (6, 1) (6, 2) (6, 3) (6, 4) (6, 5) (6, 6)$$

The number (n) of ordered pairs in the above set is:

$$n(S) = 36$$

The number (n) of ordered pairs that add up to 2 or 3 are:
Even $A = (1, 1) (1, 2) (2, 1)$
or $n(A) = 3$

The number of ordered pairs that total more than 8 are:
Event $B = (6, 3) (6, 4) (6, 5) (6, 6) (5, 4) (5, 5) (5, 6) (4, 5) (4, 6) (3, 6)$
or $n(B) = 10$

The probability for Event *A* is the number of ordered sets in Event *A* divided by the number of ordered sets in the *sample space.*

$$P(A) = \frac{n(A)}{n(S)} = \frac{3}{36} = \frac{1}{12}$$

The probability for Event *B* is the number of ordered sets in Event *B* divided by the number of ordered sets in the *sample space.*

$$P(A) = \frac{n(B)}{n(S)} = \frac{10}{36} = \frac{5}{18}$$

PRACTICE PROBLEMS: PROBABILITIES

1. Two coins are tossed. Find the probability that both show heads.

2. Two dice are rolled. What is the probability that the sum will be 7?

3. One marble is drawn from a bag containing two red, two black, and two green marbles. What is the probability that a red marble will be drawn?

ANSWERS TO PRACTICE PROBLEMS: PROBABILITIES

1. $S = (H, H) (H, T) (T, H) (T, T)$

Event $A = (H, H)$

$$P(A) = \frac{n(A)}{n(S)} = \frac{1}{4}$$

The probability is $\frac{1}{4}$ that both coins will turn up heads.

2. $S = (1, 1) (1, 2) (1, 3) (1, 4) (1, 5) (1, 6) (2, 1) (2, 2) (2, 3) (2, 4) (2, 5) (2, 6) (3, 1) (3, 2) (3, 3) (3, 4) (3, 5) (3, 6) (4, 1) (4, 2) (4, 3) (4, 4) (4, 5) (4, 6) (5, 1) (5, 2) (5, 3) (5, 4) (5, 5) (5, 6) (6, 1) (6, 2) (6, 3) (6, 4) (6, 5) (6, 6)$

The number (n) of ordered pairs in the above set is:

$$n(S) = 36$$

The number (n) of ordered pairs that add up to 7 are:

Event $A = (1, 6) (2, 5) (3, 4) (4, 3) (5, 2) (6, 1)$

or $n(A) = 6$

Divide the number of ordered sets in Event A by the number of ordered sets in the Sample Space:

$$P(A) = \frac{n(A)}{n(S)} = \frac{6}{36} = \frac{1}{6}$$

3. $S = $ (r) (r) (b) (b) (g) (g)

Event $A = $ (r) (r)

Event $A = 2$

$$P(A) = \frac{n(A)}{n(S)} = \frac{2}{6} = \frac{1}{3}$$

PERMUTATIONS AND COMBINATIONS

Although probabilities are imprecise, *the number of possible outcomes* is precise.

EXAMPLE 1:

If you have four sweaters and two pairs of jeans, how many different sweater and jean combinations can you make?

Let A, B, C, and D represent the sweaters and 1 and 2 the pairs of jeans. We can represent these 8 combinations on the diagram below.

Since there are four sweaters and two pairs of jeans from which to choose, there are $4 \times 2 = 8$ possible combinations.

In other words, if you can select one item in "X" different ways and then another selection in "Y" different ways, the number of ways the two selections can be made is $X \times Y$ different ways.

EXAMPLE 2:

What are the possible outcomes if three coins, a nickel, dime and quarter, are tossed?

H = heads

T = tails

	5	**10**	**25**
1.	H	H	H
2.	H	H	T
3.	H	T	H
4.	H	T	T
5.	T	H	H
6.	T	H	T
7.	T	T	H
8.	T	T	T

As you can see by the chart, there are eight different possible outcomes. Since each coin can turn up two possible ways, then:

$$2 \times 2 \times 2 = 8$$

EXAMPLE 3:

In an election, there are 3 candidates for senator and 4 for governor. How many ways can a ballot be marked for both of these offices?

This problem involves finding the number of different ways two acts can be performed in succession. If one action can be done in any one of n different ways, and if, after it has been done, a second action can be done in any one of m different ways, then both acts can be performed in the order stated in nm different ways.

The ballot can be marked for senator in any one of (three) ways. With any one of these (three) ways, we can associate any one of (four) ways. Thus, the number of ways a ballot can be marked for both offices is $3 \times 4 = 12$ different ways.

If we wanted to know how many ways three ballots could be marked for both offices, we would multiply each ballot times the number of possible combinations.

EXAMPLE 4:

There are four roads from City 1 to City 2 and five roads from City 2 to City 3. How many routes can you take from City 1 via City 2 to City 3?

The question deals with the number of ways each element of one set (A, B, C) can be matched with each element of another set (E, F, G, H, I).

Ordered arrangements of numbers are called PERMUTATIONS. Subsets of these arrangements are called COMBINATIONS.

Routes from City 1 to City 2: A, B, C, D

Routes from City 2 to City 3: E, F, G, H, I

EXAMPLE 5:

How many 3-digit, positive, even integers are there?

The number is even. Therefore, it must end with either a 0, 2, 4, 6 or 8. There are five different choices for the last digit.

$$\underline{\quad\quad}\ \underline{\quad\quad}\ \underline{5}$$

For the number to be meaningful, it must start with either a 1, 2, 3, 4, 5, 6, 7, 8 or 9.

$$\underline{9}\ \underline{\quad\quad}\ \underline{5}$$

The second digit may be any of the digits 0, 1, 2, 3, 4, 5, 6, 7, 8 or 9. There are ten possible numbers for the second digit.

$$\underline{9}\ \underline{10}\ \underline{5}$$

There are $9 \times 10 \times 5$ or 450 different 3-digit positive, even integers.

PRACTICE PROBLEMS

1. If a girl has six different skirts and ten different blouses, how many different skirt-and-blouse outfits can she put together?

2. A drawer contains ten spoons, six knives, and five forks. In how many ways can you choose a place setting of a knife, fork and spoon?

3. In how many ways can you answer ten true-false questions?

4. How many two-or three-digit numbers can be formed from the digits 1, 2, 3, 4 and 5, if there are no repeated digits in the number?

ANSWERS TO PRACTICE PROBLEMS

1. 6 skirts × 10 blouses = 60 different combinations.

2. $10 \times 6 \times 5 = 300$

3. 10 questions for two ways:

$$2 \times 2 \times 2 \times 2 \times 2 \times 2 \times 2 \times 2 \times 2 \times 2 \text{ or}$$

$$2^{10} = 1,024$$

4. If one of the five numbers is the first digit of a 3-digit number, then only four are left for the second digit and three for the third:

$$5 \times 4 \times 3 = 60$$

If one of the five numbers is used in the two-digit number, then only four are left for the second digit.

$$5 \times 4 = 20$$

$$5 \times 4 \times 3 = 60$$

$$5 \times 4 = 20$$

$20 + 60 = 80$ two- and three-digit numbers can be formed from the numbers 1, 2, 3, 4 and 5.

PERMUTATIONS

Each arrangement or ordering of a set of objects is called a *permutation* of the set. In the set [1, 2, 3] there are six possible *permutations*.

They are: (1, 2, 3), (1, 3, 2), (2, 1, 3), (2, 3, 1), (3, 1, 2), (3, 2, 1)

There are three possible choices for the first position, either 1, 2, or 3. But once that position is filled, there are only two choices left for the second position, and once that position is filled, only one choice left for the final position, so that:

$$P \text{ (permutations)} = 3 \times 2 \times 1 = 6$$

In other words, the number of permutations possible in a set of n objects is $n(n - 1) (n - 2) (n - 3)$, etc., or *all the whole numbers from one up to that number*.

To indicate that we want to multiply together all the numbers up to an *n* number, we write the number with an exclamation sign after it.

EXAMPLE:

$$7! = 7 \times 6 \times 5 \times 4 \times 3 \times 2 \times 1$$

So, when we see 7!, we know it represents the product of all the numbers from one to seven. This representation is called the *factorial function*.

The *factorial* of a whole number *n* is *n*! The *factorial* 7! is 5,040.

$$7! = 7 \times 6 \times 5 \times 4 \times 3 \times 2 \times 1 = 5,040.$$

STUDY PROBLEM 3

How many ways can eight people line up in a cafeteria line?

If the first person is in the first position, then only 7 people are left to line up in the second position, 6 people in the third position, etc.

$$P = 8!$$

$$8 \times 7 \times 6 \times 5 \times 4 \times 3 \times 2 \times 1 = 40,320$$

EXERCISE

Solve the following:

1. $5! =$

2. $(3!)\,(4!) =$

3. $\dfrac{7!}{5!} =$

4. How many ways can the letters in the word SUNDAY be arranged?

ANSWERS
---

1. $5 \times 4 \times 3 \times 2 \times 1 = 120$

2. $(3 \times 2 \times 1)\,(4 \times 3 \times 2 \times 1) = (6)\,(24) = 144$

3. $\dfrac{7 \times 6 \times \cancel{5} \times \cancel{4} \times \cancel{3} \times \cancel{2} \times \cancel{1}}{\cancel{5} \times \cancel{4} \times \cancel{3} \times \cancel{2} \times \cancel{1}} = 42$

4. $6 \times 5 \times 4 \times 3 \times 2 \times 1 = 720$

In Example 1 concerning sweaters and jeans, we are not interested in the order in which the sweater and jean combinations are chosen, but only in the number of possible ways they can be chosen.

However, in some situations involving choices, the order in which the choices are made *IS* important.

STUDY PROBLEM 4

Six books are selected from a group of ten and arranged on a shelf. How many arrangements are possible?

Solution: Find the number of permutations of ten taken six at a time.

The symbol *nPr* is used to indicate the number of permutations of *n* objects taken *r* at a time.

$$_{10}P_6$$

The factorial of $10 = 10 \times 9 \times 7 \times 6 \times 5 \times 4 \times 3 \times 2 \times 1$ but we will select only six, so:

$$_{10}P_6 = 10 \times 9 \times 8 \times 7 \times 6 \times 5 = 151,200 \text{ } possible \text{ } arrangements.$$

STUDY PROBLEM 5

How many different arrangements (permutations) that are distinguishable can be formed from the word LITTLE using all the letters?

There are 6! number of permutations. But since there are two T's and two L's which are repeated and not distinguishable from each other, we must divide by 2! (L's) and 2! (T's).

$$P\frac{6!}{2!2!} = \frac{6 \bullet 5 \bullet 4 \bullet 3 \bullet 2 \bullet 1}{(2 \bullet 1)(2 \bullet 1)} = \frac{720}{4} = 180$$

The number of permutations, P, of n elements, taken n at a time, with n_1 elements alike, and $n2$ of another kind alike is:

$$\frac{n!}{n_1! \, n_2!}$$

EXERCISE

Find the number of permutations of all the letters in each word.

1. Deeded

2. Cincinnati

3. How many different permutations can be made by flying six pennants at a time on a vertical flag pole, if two are red, three are blue, and one is white?

ANSWERS

---

1. $\dfrac{6!}{3!3!} = \dfrac{6 \bullet 5 \bullet 4 \bullet \cancel{3} \bullet \cancel{2} \bullet \cancel{1}}{3 \bullet 2 \bullet 1 \bullet \cancel{3} \bullet \cancel{2} \bullet \cancel{1}} = \dfrac{6 \bullet 5 \bullet 4}{3 \bullet 2 \bullet 1} = \dfrac{120}{6} = 20$

2. $\dfrac{10!}{2!3!3!} = \dfrac{10 \bullet 9 \bullet 8 \bullet 7 \bullet 6 \bullet 5 \bullet 4 \bullet \cancel{3} \bullet \cancel{2} \bullet \cancel{1}}{2 \bullet 1 \bullet 3 \bullet 2 \bullet 1 \bullet \cancel{3} \bullet \cancel{2} \bullet \cancel{1}} = \dfrac{604,800}{12} = 50,400$

3. $\dfrac{6!}{2!3!} = \dfrac{6 \bullet 5 \bullet 4}{2} = 60$

COMBINATIONS

Arrangements without consideration of the order of elements are called *combinations*. There are six permutations of three letters each using the letters a, b, and c:

$$P = (a, b, c)\,(a, c, b)\,(b, a, c)\,(b, c, a)\,(c, b, a)\,(c, a, b)$$

But, there is just one combination (a, b, c).

The combination of two elements from one set $[a, b, c]$ are:

$$(a, b)$$

$$(b, c)$$

$$(a, c)$$

The symbol C indicates this is a combination. The combinations can be represented as:

$$_3C_2$$

That is, the combination of two elements from a set of three elements.

From each combination, two permutations can be formed, or 2^P2.

Combinations	Permutations
(a, b)	$(a, b)\,(b, a)$
(b, c)	$(b, c)\,(c, b)$
(a, c)	$(a, c)\,(c, a)$

EXAMPLE:

A club of 12 members wishes to choose a president, vice president, and secretary. The order in which the members' names are written on a ballot will determine who will fill each office. How many different selections are possible?

The question deals with the number of ways each element of one set:

SET (Club Members)

1, 2, 3, 4, 5, 6, 7, 8, 9, 10, 11, 12

can be matched with each element of the subset:

SUBSET (Officers)

1 (president), 2 (vice president), 3 (secretary)

There are twelve choices for president. But once a member has been listed on a ballot as president, he cannot appear on the same ballot as vice president. The order of selection is important. Once a member has been written in as president, there are only eleven possibilities left for vice president. Once members have been chosen as president and vice president, there are only ten possibilities left for secretary.

The number of possible selections (permutations) are:

$$_{12}P3 = 12 \bullet 11 \bullet 10 = 1{,}320$$

In a second situation, 3 committee members are to be selected from a club of 12 members. The order of selection is *not important*. The selection is called a combination. The 12 indicates the number of people and the 3 indicates the number of people to be selected.

A combination of n elements taken r at a time is written as nCr.

$$nCr \text{ is defined as } \frac{n}{r!(n-r)!}$$

That is:

1. The factorial of the total elements

2. Divided by the factorial of the elements taken at one time

3. Multiplied by the factorial of the number left over

A combination of twelve elements taken three at a time is:

$$_{12}C_3 = \frac{12!}{3!(12-3)!} = \frac{12 \bullet 11 \bullet 10 \bullet \cancel{9} \bullet \cancel{8} \bullet \cancel{7} \bullet \cancel{6} \bullet \cancel{5} \bullet \cancel{4} \bullet \cancel{3} \bullet \cancel{2} \bullet \cancel{1}}{3 \bullet 2 \bullet 1 \bullet \cancel{9} \bullet \cancel{8} \bullet \cancel{7} \bullet \cancel{6} \bullet \cancel{5} \bullet \cancel{4} \bullet \cancel{3} \bullet \cancel{2} \bullet \cancel{1}}$$

The like numbers in the denominator cancel like numbers in the numerator and we are left with:

$$_{12}C_3 = \frac{12 \bullet 11 \bullet 10}{3 \bullet 2 \bullet 1} = 220$$

EXERCISES WITH COMBINATIONS

Solve the following:

1. $_7C_4 =$

2. $_4C_3 =$

3. $_{12}C_2 =$

4. $_{12}C_7 =$

5. How many ways can three prizes be awarded to a group of 8 people?

ANSWERS

1. $_7C_4 = \dfrac{7!}{4!3!} = \dfrac{7 \bullet 6 \bullet 5 \bullet 4 \bullet \cancel{3} \bullet \cancel{2} \bullet \cancel{1}}{4 \bullet 3 \bullet 2 \bullet 1 \bullet \cancel{3} \bullet \cancel{2} \bullet \cancel{1}} = \dfrac{840}{24} = 35$

2. $_4C_3 = \dfrac{4!}{4!1!} = \dfrac{4 \bullet 3 \bullet 2 \bullet \cancel{1}}{4 \bullet 3 \bullet 2 \bullet 1 \bullet \cancel{1}} = \dfrac{24}{24} = 1$

3. $_{12}C_2 = \dfrac{12!}{2!10!} = \dfrac{12 \bullet 11 \bullet \cancel{10} \bullet \cancel{9} \bullet \cancel{8} \bullet \cancel{7} \bullet \cancel{6} \bullet \cancel{5} \bullet \cancel{4} \bullet \cancel{3} \bullet \cancel{2} \bullet \cancel{1}}{2 \bullet 1 \bullet \cancel{10} \bullet \cancel{9} \bullet \cancel{8} \bullet \cancel{7} \bullet \cancel{6} \bullet \cancel{5} \bullet \cancel{4} \bullet \cancel{3} \bullet \cancel{2} \bullet \cancel{1}} = \dfrac{132}{2} = 66$

4. $_{12}C_7 = \dfrac{12!}{7!5!} = \dfrac{12 \bullet 11 \bullet 10 \bullet 9 \bullet 8 \bullet \cancel{7} \bullet \cancel{6} \bullet \cancel{5} \bullet \cancel{4} \bullet \cancel{3} \bullet \cancel{2} \bullet \cancel{1}}{\cancel{7} \bullet \cancel{6} \bullet \cancel{5} \bullet \cancel{4} \bullet \cancel{3} \bullet \cancel{2} \bullet \cancel{1} \bullet 5 \bullet 4 \bullet 3 \bullet 2 \bullet 1} = \dfrac{95040}{120} = 792$

5. $_8C_3 = \dfrac{8!}{3!5!} = \dfrac{8 \bullet 7 \bullet 6 \bullet \cancel{5} \bullet \cancel{4} \bullet \cancel{3} \bullet \cancel{2} \bullet \cancel{1}}{3 \bullet 2 \bullet 1 \bullet \cancel{5} \bullet \cancel{4} \bullet \cancel{3} \bullet \cancel{2} \bullet \cancel{1}} = \dfrac{336}{6} = 56$

PRACTICE PROBLEMS:
PROBABILITIES, COMBINATIONS, AND PERMUTATIONS

1. In a certain registration district, 18 boys and 21 girls were born during one month. Find the probability that the first child born during that month was a boy.

2. The probability that Washington High will win tonight's basketball game is $\frac{2}{5}$ and the probability that Kennedy High will win their game tonight is $\frac{3}{7}$. Find the probability that both Washington and Kennedy will win tonight.

3. John and Jose are running in different races. The probability that John will win his race is $\frac{1}{2}$. The probability that Jose will win his race is $\frac{1}{5}$. What is the probability that both will win?

4. If 10 runners compete in a race, in how many different ways can prizes be awarded for first, second, and third places?

5. In a school election, there are 2 candidates for president, 3 for vice president, 4 for secretary, and 1 for treasurer. How many ways may the election result?

6. On three shelves in the library, there are 15 books, 14 books, and 20 books, respectively. In how many ways may you choose one book from each shelf?

7. A basketball coach must choose a center, left guard, right guard, left forward, and a right forward from a team of ten players. How many ways can this be done?

8. How many pairs of letters are possible if each pair contains one of the vowels a, i, and u, and one of the consonants b, c, d, f, and g?

9. A man can choose one of four envelopes. One envelope contains $100 and the others are empty. What is the probability that he will choose the $100 envelope?

10. How many distinguishable permutations can be made from the word Mississippi?

11. How many positive odd integers less than 1,000 can be formed using the digits 2, 3, 4, 5, 6, and 8?

12. One marble is drawn from a bag of three black, four white, and six red marbles. Find the probability that the marble is black.

13. How many different permutations can be made by flying six pennants at a time on a vertical flag pole if they are all different colors?

14. Find the number of committees of 4 persons that can be selected from a set of 15 people.

15. How many ways can 5 of 8 potential jurists be seated in five vacant seats in court?

ANSWERS TO PRACTICE PROBLEMS: PROBABILITIES, PERMUTATIONS, AND COMBINATIONS

1. 18 boys + 21 girls = 39 born

$$\frac{18}{39} \text{ boys total} = \frac{6}{13}$$

2. $\frac{2}{5} \cdot \frac{3}{7} = \frac{6}{35}$

3. $\frac{1}{2} \cdot \frac{1}{5} = \frac{1}{10}$ probability both will win their races

4. $10 \cdot 9 \cdot 8 = 720$ different ways

5. $2 \times 3 \times 4 \times 1 = 24$ ways

6. $15 \times 14 \times 20 = 4,200$

7. $10 \cdot 9 \cdot 8 \cdot 7 \cdot 6 = 30,240$ ways

8. 3 vowels $\times$ 5 consonants = 15 pairs

9. $P = \frac{1}{4}$

10. $\dfrac{11!}{4!4!2!} = \dfrac{11 \cdot 10 \cdot 9 \cdot 8 \cdot 7 \cdot 6 \cdot 5 \cdot \cancel{4} \cdot \cancel{3} \cdot \cancel{2} \cdot \cancel{1}}{4 \cdot 3 \cdot 2 \cdot 1 \cdot \cancel{4} \cdot \cancel{3} \cdot \cancel{2} \cdot \cancel{1}} = \dfrac{1663200}{48} = 34,650$

11. $2 = 2$

$\underline{6} \cdot \underline{2} = 12$

$\underline{6} \cdot \underline{6} \cdot \underline{2} = \underline{72}$

86

12. $P(a) = \dfrac{n(a)}{n(s)} = \dfrac{3}{13}$

13. $6! = 720$

14. $_{15}C_4 = \dfrac{15!}{4!11!} \dfrac{15 \cdot 14 \cdot 13 \cdot 12 \cdot \cancel{11} \cdot \cancel{10} \cdot \cancel{9} \cdot \cancel{8} \cdot \cancel{7} \cdot \cancel{6} \cdot \cancel{5} \cdot \cancel{4} \cdot \cancel{3} \cdot \cancel{2} \cdot \cancel{1}}{4 \cdot 3 \cdot 2 \cdot 1 \cdot \cancel{11} \cdot \cancel{10} \cdot \cancel{9} \cdot \cancel{8} \cdot \cancel{7} \cdot \cancel{6} \cdot \cancel{5} \cdot \cancel{4} \cdot \cancel{3} \cdot \cancel{2} \cdot \cancel{1}} = $

$\dfrac{32760}{24} = 1365$

15. $_8P_5 = 8 \cdot 7 \cdot 6 \cdot 5 \cdot 4 = 6{,}720$

TABLE 1

TABLE OF SQUARES AND SQUARE ROOTS

N	N^2	$\sqrt{N}$	N	N^2	$\sqrt{N}$
1	1	1	51	2,601	7.141
2	4	1.414	52	2,704	7.211
3	9	1.732	53	3,809	7.280
4	16	2	54	2,916	7.348
5	25	2.236	55	3,025	7.416
6	36	2.449	56	3,136	7.483
7	49	2.646	57	3,249	7.550
8	64	2.828	58	3.364	7.616
9	81	3	59	3,481	7.681
10	100	3.162	60	3,600	7.746
11	121	3.317	61	3,721	7.810
12	144	3.464	62	3,844	7.874
13	169	3.606	63	3,969	7.937
14	196	3.742	64	4,096	8
15	225	3.873	65	4,225	8.062
16	256	4	66	4,356	8.124
17	289	4.123	67	4,489	8.185
18	324	4.243	68	4,624	8.246
19	361	4.359	69	4,761	8.307
20	400	4.472	70	4,900	8.367
21	441	4.583	71	5,041	8.426
22	484	4.690	72	5,184	8.485
23	529	4.796	73	5,329	8.544
24	576	4.899	74	5,476	8.602
25	625	5	75	5,625	8.660
26	676	5.099	76	5,776	8.718
27	729	5.196	77	5,929	8.775
28	784	5.292	78	6,084	8.832
29	841	5.385	79	6,241	8.888
30	900	5.477	80	6,400	8.944
31	961	5.568	81	6,561	9

32	1,024	5.657	82	6,724	9.055
33	1,089	5.745	83	6,889	9.110
34	1,156	5.831	84	7,056	9.165
35	1,225	5.916	85	7,225	9.220
36	1,296	6	86	7,396	9.274
37	1,369	6.083	87	7,569	9.327
38	1,444	6.164	88	7,744	9.381
39	1,521	6.245	89	7,921	9.434
40	1,600	6.325	90	8,100	9.487
41	1,681	6.403	91	8,281	9.539
42	1,764	6.481	92	8,464	9.592
43	1,849	6.557	93	8,649	9.644
44	1,936	6.633	94	8,836	9.695
45	2,025	6.708	95	9,025	9.747
46	2,116	6.782	96	9,216	9.798
47	2,209	6.856	97	9,409	9.849
48	2,304	6.928	98	9,604	9.899
49	2,401	7	99	9,801	9.950
50	2,500	7.071	100	10,000	10

TABLE 2
TABLE OF VALUES OF TRIGONOMETRIC FUNCTIONS

deg	sin	cos	tan	deg	sin	cos	tan	deg	sin	cos	tan
0	.000	1.000	.000	31	.515	.857	.601	61	.875	.485	1.804
1	.017	1.000	.017	32	.530	.848	.625	62	.883	.470	1.881
2	.035	.999	.035	33	.545	.839	.649	63	.891	.454	1.963
3	.052	.999	.052	34	.559	.829	.675	64	.899	.438	2.050
4	.070	.998	.070	35	.574	.819	.700	65	.906	.423	2.145
5	.087	.996	.087	36	.588	.809	.727	66	.914	.407	2.246
6	.105	.995	.105	37	.602	.799	.754	67	.921	.391	2.356
7	.122	.993	.123	38	.616	.788	.781	68	.927	.375	2.475
8	.139	.990	.141	39	.629	.777	.810	69	.934	.358	2.605
9	.156	.988	.158	40	.643	.766	.839	70	.940	.342	2.747
10	.174	.985	.176	41	.656	.755	.869	71	.946	.326	2.904
11	.191	.982	.194	42	.669	.743	.900	72	.951	.309	3.078
12	.208	.978	.213	43	.682	.731	.933	73	.956	.292	3.271
13	.225	.974	.231	44	.695	.719	.966	74	.961	.276	3.487
14	.242	.970	.249	45	.707	.707	1.000	75	.966	.259	3.732
15	.259	.966	.268								

deg	sin	cos	tan	deg	sin	cos	tan	deg	sin	cos	tan
16	.276	.961	.287	46	.719	.695	1.036	76	.970	.242	4.011
17	.292	.956	.306	47	.731	.682	1.072	77	.974	.225	4.331
18	.309	.951	.325	48	.743	.669	1.111	78	.978	.208	4.705
19	.326	.946	.344	49	.755	.656	1.150	79	.982	.191	5.145
20	.342	.940	.364	50	.766	.643	1.192	80	.985	.174	5.671
21	.358	.934	.384	51	.777	.629	1.235	81	.988	.156	6.314
22	.375	.927	.404	52	.788	.616	1.280	82	.990	.139	7.115
23	.391	.921	.424	53	.799	.602	1.327	83	.993	.122	8.144
24	.407	.914	.445	54	.809	.588	1.376	84	.995	.105	9.514
25	.423	.906	.466	55	.819	.574	1.428	85	.996	.087	11.430
26	.438	.899	.488	56	.829	.559	1.483	86	.998	.070	14.301
27	.454	.891	.510	57	.839	.545	1.540	87	.999	.052	19.081
28	.470	.883	.532	58	.848	.530	1.600	88	.999	.035	28.636
29	.485	.875	.554	59	.857	.515	1.664	89	1.000	.017	57.290
30	.500	.866	.577	60	.866	.500	1.732	90	1.000	.000	-

NOTES

NOTES

NOTES

NOTES

NOTES

NOTES

NOTES

NOTES

NOTES

NOTES

Decisions, Decisions.

Choosing the right college can be tough. So many schools, so little time. Fortunately, Peterson's suite of print and online college search and selection tools can take the guesswork out of this potentially overwhelming process.

Four-Year Colleges

With detailed profiles of every accredited college and university in the U.S. and Canada, this classic guide to college selection is perfect for students, parents and teachers. Features complete information on academic programs, student-faculty ratios, tuition costs, test scores and campus activities.

Two-Year Colleges

The only truly comprehensive reference, filled with information on more than 1,600 accredited public and private junior and community colleges in the U.S. and Canada. Students can compare curricula, associate degree programs, tuition costs, off-campus activities, and more.

For additional, more specific information, check out the rest of our college search and selection titles, including: *Christian Colleges & Universities, College & University Almanac, Professional Degree Programs in the Visual & Performing Arts, Colleges with Programs for Students with Learning Disabilities or Attention Deficit Disorders,* and *Competitive Colleges*.

Plus, for even more valuable information on the search and selection process, visit Peterson's College and University Information Channel where you can access CollegesWantYou℠, CollegeQuest®, and BestCollegePicks℠.

THOMSON

PETERSON'S